Rebel Streets and the Informal Economy

Street trade is a critical and highly visible component of the informal economy, linked to global systems of exchange. Yet policy responses are dismissive and evictions commonplace. Despite being progressively marginalised from public space, street traders in the global south are engaged in spatial and political battlegrounds to reclaim space, and claim *de facto* property rights over their place of work, through quiet infiltration, union power, or direct action.

This book explores 'rebel streets', the challenges faced by informal economy actors and how organised groups are seeking to reframe legal understandings to create new claims to space and urban rights. The book sets out new thinking and a conceptual framework for improved understanding of the plural relationship between *law, rights and space for the informal economy*, the contest between traditional, modernist and rights-based approaches to development, and impacts on the urban working poor. With a focus on street trading, the book seeks to reframe the legal context in which modern informal economies operate, drawing on key areas of academic enquiry and case studies of how vendors are staking claim to urban rights.

The book argues for a reconceptualisation of legal instruments to provide a rights-based framework for urban work that recognises the legitimacy of urban informal economies, the scope for collective management of urban resources and the social value of public space as a site for urban livelihoods. It will be of interest to students and scholars of geography, economics, urban studies, development studies, political studies and law.

Alison Brown is Professor of Urban Planning and International Development in the School of Geography and Planning, Cardiff University, UK. She is an urban planner, whose research focuses on urbanisation and development policy, local governance and urban law, livelihoods, the urban informal economy and post-conflict cities, and she has published widely on the informal economy and rights-based approaches to development. She was Principal Investigator on the research project *Making Space for the Poor: Law, Rights, Regulation and Street-Trade in the 21st Century*, funded by the Economic and Social Research Council (ESRC)/Department for International Development (DFID) Joint Fund for Poverty Alleviation Research, reported in this book. She was an expert adviser to *Habitat III*, the *UN Conference on Housing and Sustainable Urban Development* (as a member of Policy Unit 1 on the *Right to the City and Cities for All*) and has written specialist development topic guides for DFID on planning for sustainable cities in the global south and livelihoods and urbanisation. She is a board member of the NGO Reall (formerly Homeless International) and planning adviser to the global network WIEGO (Women in Informal Employment: Globalizing and Organizing).

Routledge Studies in Urbanism and the City

This series offers a forum for original and innovative research that engages with key debates and concepts in the field. Titles within the series range from empirical investigations to theoretical engagements, offering international perspectives and multidisciplinary dialogues across the social sciences and humanities, from urban studies, planning, geography, geohumanities, sociology, politics, the arts, cultural studies, philosophy and literature.

For a full list of titles in this series, please visit www.routledge.com/series/RSUC

Rebel Streets and the Informal Economy

Street Trade and the Law

Edited by Alison Brown

Routledge
Taylor & Francis Group

LONDON AND NEW YORK

First published 2017 by Routledge

2 Park Square, Milton Park, Abingdon, Oxfordshire OX14 4RN
52 Vanderbilt Avenue, New York, NY 10017

Routledge is an imprint of the Taylor & Francis Group, an informa business

First issued in paperback 2018

Copyright © 2017 selection and editorial matter, Alison Brown; individual chapters, the contributors

The right of Alison Brown to be identified as the author of the editorial material, and of the authors for their individual chapters, has been asserted in accordance with sections 77 and 78 of the Copyright, Designs and Patents Act 1988.

All rights reserved. No part of this book may be reprinted or reproduced or utilised in any form or by any electronic, mechanical, or other means, now known or hereafter invented, including photocopying and recording, or in any information storage or retrieval system, without permission in writing from the publishers.

Notice:
Product or corporate names may be trademarks or registered trademarks, and are used only for identification and explanation without intent to infringe.

British Library Cataloguing in Publication Data
A catalogue record for this book is available from the British Library

Library of Congress Cataloging in Publication Data
Names: Brown, Alison (Alison Margaret Braithwaite), editor.
Title: Rebel streets and the informal economy : street trade and the law / edited by Alison Brown.
Description: Abingdon, Oxon; New York, NY : Routledge, 2017. | Series: Routledge studies in urbanism and the city
Identifiers: LCCN 2016044545 | ISBN 9781138189744 (hardback) | ISBN 9781315641461 (ebook)
Subjects: LCSH: Street vendors–Legal status, laws, etc. | Informal sector (Economics)–Law and legislation. | Peddling–Law and legislation. | Informal sector (Economics).
Classification: LCC K1022.C6 R43 2017 | DDC 343.08–dc23
LC record available at https://lccn.loc.gov/2016044545

ISBN: 978-1-138-18974-4 (hbk)
ISBN: 978-0-367-13875-2 (pbk)

Typeset in Times New Roman
by Cenveo Publisher Services

To my collaborator, colleague and dear friend, the late Professor Michal Lyons

Contents

Illustrations

Figures

Tables

Boxes

Contributors

Tulia Ackson MP is Deputy Speaker of the National Assembly of the United Republic of Tanzania, Deputy Attorney General of the United Republic of Tanzania and Past Dean and Associate Professor of the School of the Law, University of Dar es Salaam, Tanzania.

Christine Bonner was the WIEGO Organization and Representation Programme Director, 2014–16, responsible for the Law and Informality Project. She was the founding director of the worker education institute, DITSELA, in Johannesburg, South Africa.

Alison Brown is Professor of Urban Planning and International Development at the School of Geography and Planning, Cardiff University, UK.

Ibrahima Dankoco is a Professor at the Faculté des Sciences Economiques et de Gestion, Université Cheikh Anta Diop, Sénégal.

Edésio Fernandes is an urban law expert and Fellow at the Lincoln Institute of Land Policy, USA, and Development Planning Unit, University College London, UK.

Suzanne Fitzpatrick is Professor of Housing and Social Policy at the Institute for Social Policy, Housing, Environment and Real Estate, Heriot-Watt University, UK.

Ryan Goode studied for his PhD at the Department of Geography, San Diego State University, USA.

Gengzhi Huang is Assistant Professor at the Guagnzhou Institute of Geography, Guangzhou Branch, Chinese Academy of Sciences, China.

Nezar A. Kafafy is Lecturer in Urban Design, Landscape and Architecture, Faculty of Urban and Regional Planning, Cairo University, Egypt.

Annali Kristiansen is Strategic Adviser, Middle East and North Africa, Human Rights Systems at the Danish Institute for Human Rights, Denmark.

Zhigang Li is a Professor at the School of Urban Design, Wuhan University, China.

Peter Mackie is a Senior Lecturer at the School of Geography and Planning, Cardiff University, UK.

Darshini Mahadevia is a Professor and Past Dean of the Faculty of Planning and Coordinator of the Centre for Urban Equity (CUE), CEPT University, India.

Colman Msoka is a Lecturer and the Deputy Director of the Institute of Development Studies, University of Dar es Salaam, Tanzania.

Fatma Raâch is a public international law specialist and Lecturer at the Faculty of Legal, Political and Social Sciences of Tunis, University of Carthage, Tunisia.

Caroline Skinner is Senior Researcher at the African Centre for Cities, University of Cape Town, South Africa, and Urban Policies Research Director of the global advocacy network WIEGO.

Kate Swanson is an Assistant Professor of Geography at the Department of Geography, San Diego State University, USA.

Suchita Vyas was formerly a Research Associate at the Centre for Urban Equity (CUE), CEPT University, India.

Beth Watts is a Research Fellow at the Institute for Social Policy, Housing, Environment and Real Estate, School of the Built Environment, Heriot-Watt University, UK.

Desheng Xue is a Professor and Head of Department at the School of Geographical Science and Planning, Sun Yat-Sen University, China.

1 Urban informality and 'rebel streets'

Alison Brown and Peter Mackie

Introduction

When 26-year-old Tunisian street trader Mohamed Bouazizi set himself alight in December 2010, in protest at humiliation and constant harassment from police and city officials, his death inspired revolution throughout the Arab world. The protests in Avenue Habib Bourguiba and Tahrir Square unseated despotic regimes as protestors challenged political repression and economic exclusion in a universal call for justice, fairness and the rule of law (*Guardian*, 2011). The revolution raised concern across the region about the explosive political power of marginalised street traders, but, in practice, little has changed and in cities of Africa, Asia, Middle East and Latin America legislation affecting street traders remains unreformed and punitive for the poor.

This book focuses on the interface between street trade as the most visible and most controversial component of the informal economy, its relationship with urban law and the conflicting legal environment in which street trading takes place, and the rebellions of street traders in challenging the law and claiming legal space. Although there is no universally accepted definition of the *informal economy*, the conceptual framework now adopted by the ILO includes both: 1) the *informal sector* referring to employment and production in unincorporated, unregistered or small enterprises, and 2) *informal employment* referring to employment without social protection – which includes own-account workers and employers in informal sector enterprises, contributing family workers, employees holding informal jobs, members of informal producer's cooperatives, and own-account workers producing goods exclusively for their household's use (ILO, 2013: 42).

The book is an outcome of the Urban Law Project, research funded under the Economic and Social Research Council (ESRC)/Department for International Development (DFID) Joint Fund for Poverty Alleviation Research,[1] supplemented by two linked British Academy small grants[2] and research funded under the National Natural Sciences Foundation of China.[3] This introduction explains the philosophy of the research, and then outlines central themes of the book: law and development, law and urban management and rebel streets. The introduction ends with an overview of the book structure.

Street trade is defined here as all non-criminal commercial activity that depends on access to urban public space including market trade, trade from fixed locations and mobile vending, while *public space* is framed by the social relations that determine its use. The concept of urban public space is adopted to mean physical space and the social relations that determine that space, including all space that is not delineated or accepted as private and where there is a degree of legitimate public or community use, encompassing both formal space in parks, squares and streets, and marginal or under-used edge space (Brown, 2006: 22).

The term 'street trade' is broad in meaning. The distinction between street trade, market trade, hawking and home-based enterprise is often blurred, as markets may encompass surrounding streets, street traders may be static or mobile and home-based enterprises may spill onto the street. There are also differences in terminology as street trading is also described as 'vending', 'petty trading', or 'hawking'. Terms are often country-specific and nuanced; for example, in India the term 'vendor' is widely used, whereas in South Africa 'trader' is more common.

The activity of street trade embraces the sale and purchase of the phenomenal range of goods and services bought on city streets. These include food and produce, new and second hand clothing, manufactured items (often imported), shoes, phone accessories, traditional herbs and medicine, and services such as hair cutting and braiding, selling phone credit and many others (Brown, 2006: 8).[4] The *street economy* is a wider concept, embracing all the commercial and business activities and workers that profit from the street, such as transport workers, porters, watchmen, small-scale manufacturers, rent collectors, landlords and many others, although the legal challenges are common. Excluded from this discussion are activities often considered socially illegitimate, including drug trading and prostitution.

Urban law encompasses the policies, legislation, decisions and practices that govern the functioning of cities and human settlements – covering land, housing, urban economies, operations of local government, the environment and citizens' rights. However, legislation affecting local economies is usually designed for the formal economy, and fails to support informal workers, often criminalising the working poor. Urban law affecting the informal economy is poorly documented and erratically applied. It is often framed at national level and implemented by municipalities, but rarely fully understood by informal workers. Bylaws regulating cart-pushers, kiosk owners, hawkers and businesses licences are often outdated, and prohibitive costs and lengthy procedures put business registration out of reach.

Research objectives

The Urban Law Project was a three-year research project on law, rights and regulation of the informal economy. The focus was on street trade as one of the most contested domains of the informal economy and affected by many strands or urban law. The research hypothesis is that the urban informal economy operates in a fragmented and plural regulatory environment, with conflicts between formal and informal regulatory systems that exacerbate risks, vulnerabilities and exclusions of the working poor, and are hugely damaging to the security and stability of their livelihoods, particularly at times of economic crisis.

Fieldwork for the research took place in cities with contrasting legal and poverty traditions: Ahmedabad, Dar es Salaam, Dakar and Durban, and the concepts were developed through related studies in Cairo, Tunis, Cusco, Quito and Guangzhou. All the authors in this book have contributed to the project. Four broad dimensions of informal work are explored by authors:

- the dynamics of the informal economy and street trading in each city;
- the plural legal and regulatory environment in which street trading operates;
- the regulation of street trading in practice, through self-management, informal landlords, local governments or other means; and
- the conflicts faced by street traders and their legal or physical claim to the streets.

Understanding and addressing the conflicts is crucial to developing an enabling, pro-poor regulatory framework.

Global context

Widespread informality has now become a structural characteristic of low-income urban economies and contributes significantly to GDP. In many cities of the developing world, the informal economy provides 60–80% of urban jobs and up to 90% of new jobs (Roy, 2005; Varley, 2013; ILO, 2013a). Street trade is one of the most visible and contested domains of the urban informal economy, and a crucial livelihood strategy for the poor and very poor; it provides a key source of new jobs, particularly for young people entering the job market and new migrants to cities, and supports significant urban-to-rural and international remittances (Chen *et al.*, 2002; Brown, 2006). In times of economic crisis its role is heightened and, as the 2008 global economic crisis demonstrated, it is a refuge for the working poor, but is also vulnerable to global market change. The downturn came at a time of intense debate over the potential for legal empowerment to drive poverty reduction in the heterogeneous socio-political and legal context of cities of the south (Fernandes and Varley, 1998; McAuslan, 2002). Yet policy consistently ignores the existing and potential contribution of the informal economy to both jobs and economic output.

Far from the common perception that street trade is survivalist – an outlet for local produce or manufacture – traders are now inextricably linked to global systems of exchange (Lyons *et al.*, 2008; WIEGO, 2016). Cross and Morales suggest that the modernist vision that has shaped cities in recent years has seen many attempts to ban and over-regulate street trading as a sign of 'disorder' and poverty, yet street trading has survived and thrived as 'street merchants have not simply returned to a romanticized past but created reasoned reactions to local manifestations of today's economic, cultural and social world' (Cross and Morales, 2007: 7). In contrast, de Soto (2000) argues that it is over-regulation that drives entrepreneurs to avoid regulatory bonds.

Despite extensive academic research on illegal cities (McAuslan, 1998), debates have focused on housing and land tenure (Durand-Lasserve and Selod, 2007). Discussion of law and regulation for the economies of the urban poor – usually the informal economy – has been limited, with emphasis on formalising labour and business rights (for example, ILO, 2013b) or on urban management (Chen *et al.*, 2002). Yet fieldwork consistently shows that insecurity and harassment are crucial factors undermining urban livelihoods, with forced evictions, often pursued for political or commercial ends, legitimised by draconian or outdated legislation or externally led reforms (Brown, 2006; Brown *et al.*, 2010).

Urban law affecting street trading is complex, poorly documented and erratically applied. Interpreted and implemented by municipalities, it is often framed at national level with roots in international influences (Lyons and Brown, 2009). It is rarely understood by street traders; bylaws regulating cart-pushers, kiosk owners, hawkers and businesses licences, are often colonial relics. The legal context affecting street trade can include:

- constitutional frameworks;
- policing, local government and public order law;
- highways and urban planning legislation;
- bylaws and business licensing regulations;
- public health, markets and food hygiene regulations;
- hawking and vending regulations.

The result is that trade is often illegal in multiple ways: prohibitive costs and lengthy procedures put business registration out of reach, and lack of property rights makes traders vulnerable to evictions. Instead, street trade is regulated by a panoply of informal actors including private landlords, religious or ethnic groups, market or welfare associations, unions, savings groups, the police or vigilantes, with extortion and exploitation rife. The cumulative impacts of weak or inappropriate urban law and exploitative informal processes are poorly understood.

The risks of operating in this environment are extreme. Shocks and stresses commonly include policy shifts, victimisation or exclusion of specific groups, civil unrest, police harassment, or evictions. Several authors have written powerfully about specific conflicts (for example, Potts (2007) on Operation Murambatsvina, Skinner (2008) on evictions in Durban in the run-up to international sporting events and Middleton (2003) on exclusion of street trade from Quito's historic centre), but a more general review the drivers of eviction is overdue and will be explored in this book.

Law and development

The legal systems of nation states is the backdrop against which informal activities take place. Since the end of the twentieth century, development policy has been underpinned by a global call for the rule of law which relies almost exclusively on formal legal and judicial systems, pursued vigorously by international agencies and non-governmental organisations (NGOs) (Santos, 2006). The approach relies heavily on the neoliberal development model which emphasises the importance of market-based economic growth for poverty reduction, and a judicial framework that clarifies property rights and contracts, with little consideration of the plurality of unofficial governance mechanisms that have existed for many years (Santos, 1997, 2006). Urban policies and legislation, changes to planning and land laws, the 'modernisation' of legal and institutional frameworks take little account of the diversity of provision and informal regulation actually taking place (Lyons and Brown, 2009).

Strengthening livelihoods makes a major contribution to poverty reduction, but the impact of urban law and informal processes underpinning rights to work, to space, or to representation is not well researched. For example, enshrining a 'right to work' in legislation can form an important bargaining platform in negotiations over evictions (Brown, 2009). This book argues that urban development resulting from the livelihoods of the poor often take place outside formal law, but are nevertheless based on legitimate and socially accepted processes, as explored through the chapters in this book. Sometimes these invoke formal legal processes, but often operate in tension with the law. The case studies span different continents and legal systems, to explore how street traders claim urban space and negotiate the relationship between official and unofficial legal systems through avoidance, organisation or direct confrontation.

Five core areas of academic enquiry relating to urban law and development underpin the chapters in this book. The first draws on the concept of *legal pluralism* – defined as the coexistence of multiple legal systems in a bounded physical or social space (Merry, 1988). Urban researchers now challenge the idea that the law is a neutral instrument of change, and explore power relationships between official and popular systems of justice (McAuslan, 1998: 19; Fernandes and Varley, 1998: 9). In practice, it is argued that the heterogeneous state of the developing world juxtaposes the official/unofficial, formal/informal and traditional/modern, creating a plurality of legal orders and conceptions of modernity (Fernandes, 2009; Santos, 2006) (see Chapter 2).

The second explores *human rights* as a core strand of enquiry for the research. The Universal Declaration of Human Rights (1948) is the cornerstone of international human rights law and its subsequent covenants and conventions. Balancing legitimate rights of the state, groups and individuals is crucial and complex, and some argue that the perceived need for a human rights framework is itself a response to the dramatic political and economic changes of the twentieth century and the exclusion that resulted (Freeman, 2002). The right to adequate housing and right to water are inherent to the adopted human rights instruments (Brown and Kristiansen, 2008); however, only recently has there been an interest in applying rights-based thinking to street trade – for example, by the Commission for the Legal Empowerment of the Poor (CLEP), which promoted access to justice, property rights, labour rights and business rights, emphasising the needs of small-scale, sole-trader, or own-account entrepreneurs (UNDP, 2008 a, b). The report was influenced by de Soto's ideas of legalising 'extra-legal' agreements and assets (see Chapter 3).

The third area of enquiry is based on the *right to the city* as a paradigm for urban inclusion, the concept developed by Henri Lefebvre which challenges the social and political capitalist world order, arguing that the 'use value' of city life is destroyed by the 'exchange values' imposed by industrialisation and the commodification of urban assets (Lefebvre, 1968: 67; Mitchell, 2003; Purcell, 2002; Harvey, 2003). The agenda has influenced legal practice in Brazil – where the radical City Statute, 2001, recognised the social dimension of land ownership (Rolnik and Saule, 2001) – and Ecuador where it is enshrined in the 2008 Constitution approved by national referendum. Social movements, particularly the Global Platform on the Right to the City, argue for a new universal urban paradigm on the right to the city, defined as: 'the right of all inhabitants, present and future, permanent and temporary, to use, occupy and produce just, inclusive and sustainable cities, defined as a common good, essential to a full and decent life' (GPR2C, 2016).

The fourth area explores issues of *rights to access public space*. Public space is the common ground where people carry out the functional and ritual activities that bind a community, but is also a crucial asset for the working poor (Brown, 2006: 18). The term 'public space' as used by urban planners is rather narrowly defined to include the streets, squares, plazas and parks designated as open space, but this definition excludes significant areas of space important to those in the informal economy, including space between buildings, vacant sites, or roadside verges. Battles over this public space frequently result in the exclusion and eviction of street traders (and other so-called undesirables), a process that has been described as urban revanchism (Smith, 1996; Mackie *et al.*, 2014) whereby urban space is claimed by elites to the exclusion of the poor. And yet, in some cities of the global south a more tolerant post-revanchist approach is now emerging (see Chapter 4).

The fifth area of enquiry focuses on *land and property rights*: in a civil law context the city is viewed as a set of privately owned plots with little scope for state intervention; where public interest values are recognised, zoning and compulsory purchase laws are used as instruments of state control (Fernandes and Varley, 1998: 8). Tenure systems in the south are highly complex, with multiple legal traditions and frameworks, including private tenure, public ownership of land in communist-influenced states and customary stewardship systems (Payne, 2002: 5; Durand-Lasserve, 1998). In much of Africa communal and customary land rights are widely accepted, especially where formal systems have failed to provide access to housing or services for the urban poor – in some cities as much as 70% of housing is informal (Payne, 2002). De Soto (2000) has argued that, without a clear system of property rights, the poor cannot make productive use of their assets;

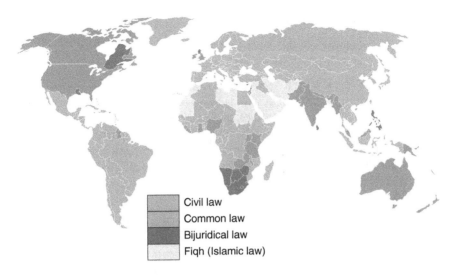

Civil law
Common law
Bijuridical law
Fiqh (Islamic law)

Figure 1.1 Legal systems of the world.
Source: Commons, 2016, Legal systems of the world, https://commons.wikimedia.org/wiki/
File:LegalSystemsOfTheWorldMap.png, accessed December 2016

despite much critique, his approach has been very influential. In general, the under-standing of property entitlements in the informal economy, the formal and informal systems that confer rights and entitlements and the mediation of disputes has received much less attention than the study of illegal settlements (see Chapter 5).

Urban law and urban management

A crosscutting theme within the book is the importance of *urban law* and the *role of local governments* in its implementation. Formal legal systems fall into various legal families, a historical legacy that influences the operation of modern-day legislation. Two main legal families are identified: civil law deriving from conti-nental Europe, which codifies the law in statutes and establishes the state as the ultimate law-maker; and English common law derived from decisions by the judiciary, which protect the individual against the state (Figure 1.1). Legal processes within the two systems are different – civil law is based on inquisition and English common law is an adversarial justice system. Box 1.1 summaries the legal systems of countries covered in Part II of this book. In a study of the effec-tiveness of African legal systems in preventing conflict, Joierman (2001) found that common law countries seem better at protecting the rule of law than civil law systems, which have a heavy toll of bureaucracy; the former may thus have more flexibility to accommodate street trading.

Local governments have a key role in implementing urban law, delivering services and tackling urban poverty, and are often the agency that translates national human

.Box 1.1 Legal traditions

This box summarises the legal systems of countries studied for this research.

India: India has a federal system of government of 28 states and seven Union territories. The legal system is based on English common law. The Constitution was adopted in 1949, and has been amended since. Part III of the Constitution establishes Fundamental Rights, which include the right to life and liberty and freedom to practice a profession, and Part IV sets out the Directive Principles of state policy. Local government is a state function, so legislation affecting street vendors varies between states.

South Africa: South Africa has a mixed legal system of Roman-Dutch civil law, English common law and customary law. Its Constitution of 1996 is the supreme law of the country. Judicial authority in South Africa is vested in the courts, with a hierarchy of jurisdictions. Traditional courts are recognised in some rural areas. Street trade is broadly regulated under company law and urban bylaws.

China: The legal system of the People's Republic of China is based primarily on the model of civil law, influenced by Soviet and continental European systems, officially defined as a 'socialist legal system'. Laws are promulgated by the National People's Congress (NPC), Administrative Regulations by the State Council, Local Regulations by the local People's Congress and local rules by local governments.

Sénégal: Sénégal's legal system is based on French civil law, with judicial review of legislative acts in a Constitutional Court. The Constitution was first adopted in 1963 and the fourth Constitution was adopted by referendum in 2001. Article 25 states that everyone has the right to work and seek employment, and a worker may join a union. Customary land tenure is recognised in the 1964 land law.

Tanzania: Tanzania's legal system is based on English common law, with limited judicial review of legislative instruments. The Constitution was adopted in 1977, with revisions in 1984 and a Bill of Rights in 1988. The government promotes economic, political and social reform, focusing on expanding the market economy, strengthening human rights, promoting democracy and environmental protection. Customary land tenure is enshrined in the 1999 Land Act and 1999 Village Land Act.

Tunisia: Tunisia is a civil law republic; the legal system is based on French civil law and Islamic law, based mainly on a series of codes, laws, decrees and ministerial orders. A unified system of legislation is applied regardless of religion. The Constitution was adopted in 1959 and amended in 1988. The post-revolution Constitution was adopted in January 2014.

Egypt: Egypt has a mixed legal system based on Napoleonic civil law, particularly French codes, and Islamic law. The Egyptian Civil Code, No. 131 of 1948, is the most important source of law in Egypt. Sharia courts were integrated into the national court system in 1956. Egypt's second post-revolution Constitution was passed by referendum in 2014.

Hertel, 2009; Mahadevia *et al.*, 2012; CIA, 2016;
Telelaws, 2016; SAG, 2016

rights strategies into practical action. Local governments have successfully adopted various human rights mechanisms to strengthen inclusion in local government policy. Of particular note is the concept of 'human rights in the city', developed by United Cities and Local Government through their *European Charter for the Safeguarding of Human Rights in the City*, and the *Global Charter Agenda for Human Rights in the City*, which makes specific reference to urban livelihoods (HRC, 2015)

However, city-level politics and processes are often complex and opaque, and city governments may be heavily constrained by limited jurisdictions, fragmented responsibilities, conflicts with higher tiers of government, limited finance and weak capacity (Devas 2004: 192–3). Day-to-day management of the informal economy also depends on the mix of formal regulations and bylaws, and informal social control by a range of urban actors, including local government officials. Rights to trade and rights to trading space are tenuous and often challenged (Fafchamps and Minten, 2001). For example, in Nairobi, bylaws cover solid waste management, hawking, *matatus*, food shops, business licensing, hand carts, etc. (Nairobi City Council, 2013).

A wide range of local government policies, regulations and bylaws affect street traders (for example, Chen *et al.*, 2002; Brown, 2006: 176–80). These include: land use and zoning; property, use and access rights to space; public health standards; business registration and operations; basic infrastructure provision – for example, electricity, water, toilets, solid waste management; and municipal market regulation. City governments can contribute to poverty reduction by ensuring secure tenure for trading, but, in practice, their actions often favour the powerful to the detriment of the poor (Devas, 2004: 190; Brown and Lloyd-Jones, 2002). *Ad hoc* negotiations between municipalities and traders over spaces of tolerance and semi-formal taxation were once common, but traders have consistently lost ground, and public space has become a battle-ground (Skinner, 2008; Brown *et al.*, 2010; Mackie *et al.*, 2014).

Some administrations have specific street trader and hawker bylaws or regulations. A study undertaken in 2011 for the National Association of Street Vendors of

Box 1.2 Comparison of street trader regulations

Licensing

Yale Law School (2011) found four general approaches to the assignment of street vending rights: 1) licensing, 2) ownership, 3) organisation-based representation, and 4) no registration. Several aspects were important.

- *Criteria for issuing licenses*: these often require information about the trader, the goods sold and location of sale.
- *Discretion in licensing*: strict rules on license allocation promoted transparency, but some discretion helped individual traders.
- *Assignment, subletting and resale*: preventing subletting was usually considered fair when trading space was scare, except where assistants were employed or there was a death of a vendor.
- *Fixed versus mobile vending sites*: some jurisdictions gave different licenses for fixed or mobile vending, restricting vendors to one permit.
- *License quotas*: some jurisdictions tried to limit the total number of licenses issued, but these were considered overly restrictive.
- *License allocation and preference schemes*: three approaches were: i) priority for disadvantaged groups; ii) a lottery; iii) a queueing system.

Identifying vending sites

- *Selecting trading sites*: sites may be i) identified in regulations; ii) selected by an urban authority; or iii) allocated in urban plans. Best practice involved traders in site selection.
- *Relocation to off-street markets*: several jurisdictions promoted such relocations, but these were problematic due to high rents, limited pitches, and lack of custom, and vendors often returned to the street.
- *Excluded locations and times*: these were implemented to ensure space management, protect public rights-of-way and protect the interests of neighbouring retailers and residents.

Administration and enforcement

- Rule-making body was by the local authority, special agency or trader group.
- Regulations established rights to form unions or associations, and rights of representation.
- Accountability includes notice of enforcement, tackling bribery and corruption, appeals and punishments (e.g. fines, confiscations etc.).

Yale Law School, 2011

India, who were lobbying for the introduction of national legislation to support street vending, compared street trade regulations in 23 countries in Africa, Asia, Europe, Latin America and North America, highlighting key concepts adopted in these regulations. The analysis covered a range of aspects of regulations that affect street traders, including licensing, site identification, administration, enforcement and sanctions (Yale Law School, 2011) (Box 1.2).

However, in practice, in parallel to formal mechanisms of control, a wealth of informal norms govern the management of trading communities (for example, Brown *et al.*, 2010; Tostensen *et al.*, 2001). These derive from a rich array of associational activities, including trades unions, religious groups, family and kinship groups, savings associations, burial societies and many more, which perform a wealth of functions, including social welfare, business support, micro-finance and advocacy. Of particular interest in this research is the role of informal mechanisms of securing rights. This book explores some of these processes.

Rebel streets and a right to the city

The title of this book pays respect to David Harvey's, *Rebel Cities: From the Right to the City to the Urban Revolution* (Harvey, 2012). In this he explores the potential of the collective *right to the city* as a challenge to capitalism and the commodification of urban space, services and facilities, and the resulting intensification of poverty and social exclusion.

Harvey's core thesis is that capitalism rests on the quest for profit and a surplus product, which has largely been absorbed through urbanisation and the creation of value in urban land. While this process has been evident for many years, for example in redevelopment plans for Paris in the 1860s, the recent urban development boom has gone global, underpinned by the new financial institutions that organise credit, package mortgages and sell these in a global market, resulting in building booms from Mexico City, to Mumbai, and Moscow and elsewhere, with a boom and bust as prices overheat. Transforming the quality of urban life into a commodity has resulted in incredible transformations of lifestyles for those with money, but an 'increasing polarisation in the distribution of wealth and power indelibly etched into the spatial forms of cities, which increasingly become cities of fortified fragments, of gated communities and privatised public spaces kept under constant surveillance' (ibid.: 15).

Harvey proposes a fundamental transformation of the way in which cities are created, and the social relations of citizens through the Lefebvrian claim of the right to the city. He argues that:

> The right to the city is, therefore, far more than a right of individual or group access to the resources that the city embodies: it is a right to change and reinvent the city more after our hearts' desire ... Thus, to claim the right to the city ... is to claim some kind of shaping power over the processes of

> urbanisation, over the ways in which our cities are made and remade, and to do so in a fundamental and radical way.
>
> Ibid.: 4, 5

The claim, Harvey argues, can only be made through ongoing class-based and anti-capitalist struggle because the forces of capitalism have to subvert an urban population which can never be totally controlled (ibid.: 116). The history of class-based urban struggle, he argues, is stunning, from the revolutionary movements in Paris in 1789 through to 1839, Spanish uprisings during the Civil War to the urban-based movements of 1968 in Paris, Chicago, Mexico City, Bangkok, Madrid and many others, up to the Occupy Movement that swept through 951 cities in 82 countries in 2011, when young city-dwellers occupied streets in protest at their lack of claim to city life. Each rebellion, Harvey argues, is an anti-capitalist struggle to dismantle privilege for the few in favour of benefits for the many.

This book draws on Harvey's philosophy to examine processes of rebellion and to argue for a more inclusive, just city in which workers in the urban informal economy, specifically street vendors, are seen as legitimate urban actors with a stake in local economies and a right to exist and thrive in cities. The book argues for a collective right to the city.

Street traders' struggle to claim work-space and marginal profit is relentless, and rebellion takes many forms. Sometimes sheer force of numbers can allow low-income hawkers to invade the busy pedestrian thoroughfares that make their trade viable – such tactics involve travelling light, carrying few goods and being poised to flee if the police arrive. Others use social relations for support, often working under a 'landlord' or 'gang-master' who may supply goods for sale, negotiate with the police and take a cut in the profits. Some traders are organised or unionised under a leader who can access political or local authority chiefs. Occasionally, direct action flares up, as traders riot against an injustice to occupy the streets, but too often the response is quasi-military action, with tear gas used to dispel traders and bulldozers to flatten their stalls. Only rarely, as two chapters in this book show, have organised groups of traders managed to claim their rebellion through the courts.

Law and livelihoods: book structure

The book has three sections. Part I, 'Rebel Streets': Law, Rights and Space in Urban Development, starts with six contextual chapters. Alison Brown first examines the implications for street trade of four theoretical debates: legal pluralism; legal empowerment; urban rights and urban governance reform, arguing that the plural influences that affect street trade have been largely ignored in policy processes. Beth Watts and Suzanne Fitzpatrick then examine the difference between natural and human rights, exploring the difficulties of defending programmatic rights which are not enshrined in legislation. Edésio Fernandes explores the concept of the right to the city, and its application in

Brazil through the 2001 City Statute in granting collective rights to urban land and participatory planning. Drawing on experience in Latin America, Peter Mackie, Kate Swanson and Ryan Goode draw on ideas of revanchist urbanism to discuss how street trader resistance is reclaiming urban space. Alison Brown then explores how property rights in public space are being reframed for street traders by conceptions of communal tenure and collective need. Chris Bonner, outlines the WIEGO[5] law programme, in which grassroots organisations of women workers identify and define the impact of urban law on their lives.

Part II, Street Trading at the Front Line, examines experience from eight different cities and countries. In Ahmedabad, Darshini Mahadevia and Suchita Vyas explore how women street vendors from the Self-Employed Women's Association (SEWA) took their case demanding a fairer policy environment to the courts, and the mixed outcome that ensued. Caroline Skinner examines the swings in vending policy in Durban and how street traders have used litigation to fight development proposals and challenge restrictive bylaws. In Guangzhou, Gengzhi Huang, Desheng Xue and Zhigang Li examine the ambivalent local authority approach despite and exclusionary policies. In Sénégal, Ibrahima Dankoco and Alison Brown examine how traders claimed rights through direct action, creating scope for political negotiation. Tanzania's punitive legal regime and the fluctuating policy environment towards street trading is explored by Tulia Ackson and Colman Msoka in their analysis of the political economy and legal framework affecting street trading in Dar es Salaam. Finally, two chapters – one by Annali Kristiansen, Alison Brown and Fatma Raâch and one by Nezar A. Kafafy – chart the experience of street vendors during the Arab Spring in Tunis and Cairo, and how street traders have faced mixed fortunes since the death of the Tunisian street trader Mohamed Bouazizi.

In Part III, Claiming 'Rebel Streets', the conclusion, draws core themes from the research, demonstrating that street traders negotiate complex and unsupportive urban law through a myriad of informal mechanisms, through subversive occupation of unobserved space, violent demonstration of their rights, effective organisation and legal challenge, to claim their rights as urban citizens.

Notes

1   *Making Space for the Poor: Law Rights and Regulation for Street Trade in the 21st Century*, DFID/ESRC Joint Fund for Poverty Alleviation Research, Project RES-167-25-0591.

2   British Academy Small Grants:
    i) Rights in the City: Developing Pro-poor Policies with Street-traders in Latin America and
    ii) Economic Inclusiveness in an Age of Revolution: Street Vending and Control of Public Space in Tunis and Cairo.

3   National Natural Sciences Foundation of China (Ref: 41130747; 41401169).

4   Some authors use the term 'street trader' and 'street vendor' interchangeably.

5   WIEGO (Women in Informal Employment: Globalizing and Organizing) is a global policy advocacy network working to increase the voice, visibility and viability of workers in the informal economy.

References

Brown, A. (ed.) (2006) *Contested Space: Street Trading, Public Space and Livelihoods in Developing Cities*, Rugby: ITDG.

Brown, A. (2009) Rights to the city for street traders and informal workers, in Jouve, B. (ed.), *Urban Policies and the Right to the City*, Paris: UNESCO, pul, Citurb, Presses universitaires de Lyon.

Brown, A. and Kristiansen, A. (2008) *Urban Policies and the Right to the City: Rights, Responsibilities and Citizenship*, Paris: UNESCO, UN-Habitat, http://unesdoc.unesco. org/images/0017/001780/178090e.pdf, accessed August 2009.

Brown, A. and Lloyd-Jones, T. (2002) Spatial planning, access and infrastructure, in Rakodi, C. and Lloyd-Jones, T. (eds), *Urban Livelihoods: A People-centred Approach to Reducing Poverty*, London: Earthscan: 188–204.

Brown, A., Lyons, M. and Dankoco, I. (2010) Street traders and the emerging spaces for urban voice and citizenship in African Cities, *Urban Studies*, 47(3): 666–83.

Chen, M., Jhabvala, R. and Lund, F. (2002) Supporting workers in the informal economy: a policy framework, *Working Paper on the Informal Economy*, Geneva: Employment Sector, International Labour Office.

CIA (2016) *World Factbook*, https://www.cia.gov/library/publications/the-world-factbook/, accessed August 2016.

Commons, 2016, Legal systems of the world, https://commons.wikimedia.org/wiki/ File:LegalSystemsOfTheWorldMap.png, accessed December 2016.

Cross, J. and Morales, A. (2007) *Street Entrepreneurs: People, Place and Politics in Local and Global Perspectives*, Abingdon: Routledge.

de Soto, H. (2000) *The Mystery of Capital: Why Capitalism Triumphs in the West and Fails Everywhere Else*, New York: Basic.

Devas, N. (ed.) (2004) *Urban Governance, Voice and Poverty in the Developing World*, London: Earthscan.

Durand-Lasserve, A. (1998) Law and urban change in developing countries: trends and issues, in Fernandes, E. and Varley, A., *Illegal Cities: Law and Urban Change in Developing Countries*, London: Zed.

Durand-Lasserve, A. and Selod, H. (2007) The formalisation of urban land tenure in developing countries, World Bank's 2007 Urban Research Symposium, 14–16 May, Washington DC.

Fafchamps, M. and Minten, B. (2001) Property rights in flea-market economy, *Economic Development and Cultural Change*, 49(2): 229–67.

Ferndandes, E. (2009) Law and land policy in Latin America: shifting paradigms and possibilities for action, *Land Lines*, July: 14–18.

Fernandes, E. and Varley, A. (1998) Law, the city and citizenship, in developing countries: an introduction, in Fernandes, E. and Varley, A., *Illegal Cities: Law and Urban Change in Developing Countries*, London: Zed.

Freeman, M. (2002) *Human Rights: An Interdisciplinary Approach*, Cambridge: Polity Press.

GPR2C (2016) What's the Right to the City, GPR2C (Global Platform for the Right to the City), http://www.righttothecityplatform.org.br/1790-2/, accessed June 2016.

Guardian, 2011, Mohammed Bouazizi: the dutiful son whose death changed Tunisia's fate, Thursday, 20 January.

Harvey, D. (2003) Debates and developments: the right to the city, *International Journal of Urban and Regional Research*, 27(4): 939–41.

Harvey, D. (2012) *Rebel Cities: From the Right to the City to the Urban Revolution*, London and New York: Verso.

Hertel, C. (2009) Comparative Law: legal systems of the world – an overview, *Notarius International* 1-2, http://www.notarius-international.uinl.org/DataBase/2009/Notarius_2009_01_02_hertel_en.pdf, accessed August 2016.

HRC (2015) HRC 30th, A/HRC/30/49, Session 7/08/2015, Role of local government in the promotion and protection of human rights: final report of the Human Rights Council Advisory Committee, http://ap.ohchr.org/documents/alldocs.aspx?doc_id=25480, accessed April 2016.

ILO (2013) *Measuring informality: a statistical manual on the informal sector and informal employment*, Geneva, ILO.

ILO (2013a) *Women and Men in the Informal Economy: A Statistical Picture*, 2nd edn, Geneva: International Labour Office (ILO), http://www.ilo.org/stat/Publications/WCMS_234413/lang--en/index.htm, accessed August 2015.

ILO (2013b) *Transitioning from the Informal to Formal Economy*, Geneva: ILO, http://www.ilo.org/wcmsp5/groups/public/---ed_norm/---relconf/documents/meetingdocument/wcms_218128.pdf, accessed August 2015.

Joierman, S. (2001). Inherited legal systems and effective rule of law: Africa and the colonial legacy, *Journal of Modern African Studies*, 39(4): 571–96.

Law Library of Congress (2016) Introduction to China's legal system, http://www.loc.gov/law/help/legal-research-guide/china.php, accessed April 2016.

Lefebvre, H. (1968) *Right to the City*, English translation of 1968 text in Kofman, E. and Lebas, E. (eds and trans) (1996) *Writings on Cities*, Oxford:Blackwell.

Lyons, M. and Brown, A. (2009) The death and birth of legal pluralism: street-trade and reforms in sub-Saharan Africa, paper to the Association for Heterodox Economics, Annual Conference, 8–9 July, Panel 2, Informal economy in the developing world: social networks and sustainability, Kingston University.

Lyons, M., Brown, A. and Li, Z.-G. (2008) The third tier of globalization: African traders in Guangzhou, *City*, 12(2): 196–206.

Mackie, P.K., Bromley, R.D.F. and Brown, A. (2014) Informal traders and the battlegrounds of revanchism in Cusco, Peru, *International Journal of Urban and Regional Research*, 38(5): 1884–903.

Mahadevia, D., Vyas, S., Brown, A. and Lyons, M. (2012) Law, rights and regulation for street vending in globalising Ahmedabad, http://www.cardiff.ac.uk/cplan/research/funded-projects/making-space-for-the-poor, accessed June 2013.

McAuslan, P. (1998) Urbanization, law and development: a record of research, in Fernandes, E. and Varley, A., *Illegal Cities: Law and Urban Change in Developing Countries*, London: Zed.

McAuslan, P. (2002) Tenure and the law: the legality of illegality and the illegality of legality, in Payne, G., *Land, Rights and Innovation: Improving Tenure Security for the Urban Poor*, Rugby: ITDG.

Merry, S.E. (1988) Legal pluralism, *Law and Society Review*, 22(5): 869–96.

Middleton, A. (2003) Informal traders and planners in the regeneration of historic city centres: the case of Quito, Ecuador, *Progress in Planning*, 59(2): 71–123.

Mitchell, D. (2003) *The Right to the City: Social Justice and the Fight for Public Space*, New York and London: Guilford Press.

Nairobi City Council (2013) Know your bylaws, http://www.nairobi.go.ke/assets/Documents/CITY-BYLAWS.pdf, accessed June 2016.

Payne, G. (2002) *Land, Rights and Innovation: Improving Tenure Security for the Urban Poor*, Rugby: ITDG.

Potts, D. (2007) The state and the informal in sub-Saharan African urban economies: revisiting debates on dualism, cities and fragile states, Working Paper No. 18, London: King's College.

Purcell, M. (2002) Excavating Lefebvre: the right to the city and its urban politics of the inhabitant, *Geojournal*, 58: 99–108.

Raisch, M. (2013) Religious legal systems in comparative law: a guide to introductory research, Hauser Global Law School Programme, http://www.nyulawglobal.org/globalex/Religious_Legal_Systems1.html, accessed April 2016.

Rolnik, R, and Saule N. (eds) (2001) Estatuto da Cidade – Guia para Implementação pelos Municipios e Cidadãos, Brasil, Brasília, Câmara dos Deputados, Coordenação de Publicações.

Roy, A. 2005. Urban informality: toward an epistemology of planning. *Journal of the American Planning Association*, 71(2): 147–58.

SAG (2016) South African Government: Judicial system, http://www.gov.za/about-government/judicial-system, accessed April 2016.

Santos, B. de Souza (1997) Three metaphors of law: the frontier, the baroque, and the south, *Law and Society Review*, 29(4): 569–84.

Santos, B. de Souza (2006) The heterogeneous state and legal pluralism in Mozambique, *Law and Society Review*, 40(1): 39–75.

Skinner, C. (2008) The struggle for the streets: processes of exclusion and inclusion of street traders in Durban, South Africa, *Development South Africa*, 25(2).

Smith, N. (1996) *The New Urban Frontier: Gentrification and the Revanchist City*, New York: Routledge.

Telelaws (2016) Tunisia, http://www.telelaws.com/tunisia/introduction/, accessed April 2016.

Tostensen, A., Tvedten, I. and Vaa, M. (eds) (2001) *Associational Life in African Cities: Popular Responses to the Urban Crisis*, Uppsala: Nordic Africa Institute.

UNDP (2008a) *Making the Law Work for Everyone: Volume 1, Report of the Commission on Legal Empowerment of the Poor*, New York: UNDP.

UNDP (2008b) *Making the Law Work for Everyone: Volume 2, Report of the Commission on Legal Empowerment of the Poor*, New York: UNDP.

Varley, A. (2013) Postcolonialising informality? *Environment and Planning D: Society and Space*, 31(1): 4–22.

WIEGO (2016) Informal economy monitoring study, http://wiego.org/wiego/informal-economy-monitoring-study-iems, accessed April 2016.

Yale Law School (2011) Developing national street vendor legislation in India: a comparative study of street vending regulation, Working paper: Transnational Development Clinic, Jerome N. Frank Legal Services Organization, Yale Law School, https://www.law.yale.edu/system/files/documents/pdf/Clinics/TDC_Comparative StudyStreetVending.pdf, accessed April 2016.

Part I

'Rebel streets'

Law, rights and space in urban development

2 Legal paradigms and the informal economy

Pluralism, empowerment, rights or governance?

Alison Brown

Summary

This chapter explores the implications of four theoretical domains for street trade: legal pluralism; legal empowerment; urban rights and the right to the city; and urban governance reform agendas. The core argument is that plural influences occur in different forms – in formal legal texts, in their implementation and in informal practice; these have potential for re-envisioning the operation of urban legal frameworks that affect street traders but have been largely ignored in policy processes.

Introduction

The turn of the twenty-first century was marked by profound economic, legal and social change. Economic recovery was on the horizon after the late-1990s financial crises in South-East Asia, Russia and Argentina; China's accession to the World Trade Organization in 2001 heralded a decade of global expansion; and the World Bank's high-profile *Doing Business* rankings, set up in 2003, promoted legal and regulatory change in 11 key business sectors for 89 countries. Meanwhile, debt relief under the Heavily Indebted Poor Countries (HIPC) initiative was gathering pace, launched in 1996 and affecting 39 countries, and in September 2000, with much fanfare, the United Nations launched the aspirational Millennium Development Goals to tackle global poverty by 2015.

Underpinning responses to each of these milestones was a profound belief in the power of the neoliberal development model, with a transparent legal system at its core to address issues of economic growth and poverty reduction. As Santos noted:

> Multilateral financial agencies and international aid non-governmental organisations (NGOs) made … [judicial reform] one of their priorities for their efforts in the developing world. The global nature of this process and the intensity with which it was implemented, in both financial and political terms, reflected the rise of the … neoliberal development model … [demanding] greater reliance on markets and the private sector: only when the rule of law

is widely accepted and effectively enforced are certainty and predictability guaranteed.

Santos, 2006

Yet, in all the debate on economic transformation and growth, the every-day reality of informal economy work – a significant contributor to both *jobs and the economy* – was largely ignored. In a rare statistical analysis of urban employment, Herrera *et al.* (2012) found that in all except one of the 11 cities in Africa, Asia and Latin America more than half of the population worked in the informal economy (excluding agricultural employment). In Bamako the proportion rose to 82%. In all cities, the proportion of women working in informal employment was higher than men. Estimates were drawn mainly from household surveys carried out by national statistics offices between 2001 and 2007. Thus, far from being an exception or temporary phenomenon, the informal economy remains the mainstay for urban employment in many cities.

The economic contribution of the informal economy is also significant. ILO research in 47 countries found that in sub-Saharan Africa (Benin, Burkina Faso, Cameroon, Niger, Sénégal and Togo) the contribution of the informal sector, excluding agriculture, to non-agricultural gross added value (GVA), was 36–62%, in India the figure was 46%, while in Latin America (Colombia, Guatemala, Honduras and Venezuela) the range was 16–34% (ILO, 2013a: 22).

Yet informal economy workers are often acutely vulnerable because of lack of legal protection (Brown, 2006: 198–201). The ILO, in its report *Transitioning from the Informal to Formal Economy*, highlights the severe decent work deficits faced by informal economy workers, including inadequate and unsafe working conditions, absence of collective bargaining rights and ambiguous employment status (ILO, 2013b). The report argues that to promote decent work the negative aspects of informality must be eliminated while ensuring that opportunities for livelihoods and entrepreneurship are not destroyed (ibid.: 70).

Legislation affecting livelihoods is often conflicting, outdated and punitive for urban workers. Thus, in reducing vulnerability for the working poor, it is crucial to question the foundation of legal and regulatory systems and the social and political economy orderings that these represent, which this chapter seeks to do. Two critical questions are posed: first, how can legal and regulatory systems be reframed to encompass the complexity of the informal economy practices and, second, how can informal traditions and systems that underpin trade be accommodated without victimization of the poor?

This chapter first examines four theoretical domains that provide insights into emerging debates on law and the informal economy which form the conceptual framework from which this research has been developed:

- the domain of *legal pluralism*, a long-established debate on the relationship between citizen and the state, which has evolved from initial examination of folk and indigenous law to address pluralism in modern industrialized societies;

- the *legal empowerment agenda* of the Commission for Legal Empowerment of the Poor (CLEP) promoted as a development agenda by the United Nations Development Programme (UNDP) from 2006;
- *urban rights debates* and the challenges of the 'right to the city' movement, and the work of scholars such as Henri Lefebvre and David Harvey;
- *urban governance reform* agendas in the developing world.

The chapter argues that plural influences occur in different guises – in formal legal texts, in their implementation and in informal practice – but the poverty impacts for legal reform of such complexity are largely unexplored. In addition, collective systems of order and justice and their relations with local government administrations have unexplored potential for strengthening the livelihoods of the poor.

Legal pluralism and the heterogeneous state

Since its inception in the early 1970s, the discipline of legal pluralism has moved from the study of tribal and indigenous law to the analysis of social ordering in advanced capitalist economies. The debates have significant potential for deepening understanding of the rules of engagement in informal economy work, derived from traditional custom or modern urban practice. Thus, of critical importance for the argument of this book is to examine the extent to which the informal rules and practices of informal economy operations can be absorbed within existing legal frameworks, suggest new legal instruments, or herald a fundamental reordering of urban law.

Legal pluralism has been defined as a 'situation in which two or more legal systems coexist in the same social field' and a legal system as 'the system of courts and judges supported by the state as well as non-legal forms of normative ordering', to include the written codes and tribunals of corporations or institutions and the informal social ordering of families, work groups or collectives (Merry, 1988: 870), although its definition and analytical tools are hotly debated.

Some see legal pluralism as a result of economic and social transformation. Moore (1973: 720) suggests that societies operate as a 'semi-autonomous social field', with internal rules and customs, but are vulnerable to external change. She cites the Chagga tribe of northern Tanzania who, in less than a 100 years, have changed from a society of warring chiefdoms to prosperous Swahili-speaking farmer-communities profiting from coffee-growing as a cash crop (ibid.: 730). Moore found that, despite Tanzania's post-independence socialist revolution that nationalized land and abolished chiefs, the Chagga continued to combine old and new and traditions which limits the effectiveness of new legislation as an instrument of social engineering (ibid.: 742).

Others see legal pluralism in terms of power relations between societal groups. Merry (1988) distinguishes between *classic legal pluralism*, which examined the encounter between indigenous and European law where imposed law confronted traditional systems of justice, and the *new legal pluralism*, which explores unofficial social ordering between dominant and subordinate groups. New legal pluralism, she argues, extends the field of study to societies without a colonial past and to developed countries with a dominant legal system, but plural orders are, to a greater or lesser extent, evident in virtually all societies (Benda-Beckmann, 2002).

A crucial question for legal pluralists is whether *law* is synonymous with *state law*. In a blistering critique, Tamanaha (1993) argues that if the 'legal' includes the 'non-legal', the definition of law must be independent of the state. Instead, he suggests that state law and non-state norms are two are starkly contrasting phenomena and that to run them together irresponsibly broadens the concept of law and equalizes normative orders that are fundamentally different in form, structure and sanctions (ibid., in Benda-Beckmann, 2002: 38). Legal pluralists, he argues, should accept that a state law model is the only sensible way to define law, although this raises the problem of whether law can exist where there is no effective state. In contrast, Croce (2012) argues the case that there is no normative difference between state law and customary or other kinds of law.

An ongoing problem remains the definition of non-state forms of social ordering, especially if this is not synonymous with customary law. Some argue that *customary law* is far from traditional, and is a product of colonial encounters between pre-capitalist societies and the colonial state. In Africa, colonial regimes saw the future as being based on 'modernization', foreign investment and a strong state underpinned by European conceptions of law. In a study of land disputes between the Banjul people of Lower Casamance in Sénégal and the French administration, Snyder (1981) showed how the rain priests' traditional relationships with land were misinterpreted in translation as 'mastery of land', in an attempt to rephrase African legal norms in European terms.

A legal pluralist perspective has also been used to criticize the neoliberal paradigm underpinning global growth and poverty reduction agendas. For example, in sub-Saharan Africa, by the mid-1980s, mushrooming government spending and poorly focused investment led to mounting debt and, by the early 1990s, some 46 countries in sub-Saharan Africa were borrowing from the World Bank, many adopting imposed or voluntary structural adjustment programmes, with accompanying legal reforms to slash government spending, remove internal subsidies and open borders international competition (Winters and Kulkarni, 2014). Yet Santos (1987, 2006) argues that this neoliberal development model, with its reliance on markets and the private sector, 'focused exclusively on the official legal and judicial system, ... and left out of consideration the multiplicity of unofficial legal orderings and dispute resolution mechanisms that had long coexisted with the official legal system' (Santos, 2006: 40). This duality he calls the 'heterogeneous' state, characterized by starkly different political and legal cultures in different economic and social fields. Similarly, Griffiths (2011) argues the mobile and contingent nature of law, in which international, national and local domains intersect.

Much discussion of legal pluralism has focused on the juxtaposition of state law and customary law in relation to land rights and tenure. McAuslan (2005) argues that there are several core preconditions for the success of plural systems, including: recognition and protection of communal and collective land rights; provision to opt from one system to another; local land administration; use of both statutory and customary dispute resolution mechanisms; recognition of urban 'customary' tenure in informal settlements; and participatory, community planning, not top-down planning.

Legal pluralism has not been widely used to address the informal economy, yet, despite its difficulties, it forms a useful concept for interrogating the case studies in this research, and the dichotomies of official/unofficial, formal/informal, traditional/modern and monocultural/multicultural orderings discussed by Santos. Critical questions then arise about pluralistic regulation of informal urban practice. Who are the gatekeepers, who controls access to space and who pays whom for protection or services? Finally, to what extent does the informal economy represent another arm of free market operations or a collective and social product?

Legal empowerment

More explicit recognition of the informal economy is found in the legal empowerment literature. The concept of legal empowerment stems partly from the work of Peruvian economist Hernando de Soto, who argues that people are trapped in poverty because the informality of their land, housing or small businesses prevents the realization of their true economic value, and that legislative reform is essential so that the assets and business undertakings of the poor can be integrated into the formal economy (de Soto, 1989, 2000).

In 2005 CLEP was set up under the auspices of the United National Development Programme, co-chaired by former US Secretary of State, Madeleine Albright and de Soto. The report draws on de Soto's thesis that formal, individual property rights and registered business operations are crucial to effective functioning of markets and that lack of title represents 'dead capital' that cannot be used to raise loans or build assets. Based on 22 national consultations, its report, *Making the Law Work for Everyone* in 2008 (UNDP, 2008), argues that 4 billion people around the world are excluded from the rule of law, and that access to legal processes is a crucial component of poverty reduction. The report includes some, albeit inconsistent, recognition of the informal economy.

Chapter 4 of the CLEP report argues that basic legal rights for informal businesses should include a *right to work* and *right to a work-space* (including public land and private residences) (ibid.: 201). This requires municipal bylaws that allow street traders to operate in public spaces, permit use rights to public land and provide effective management of informal businesses in crowded localities such as central business districts (CBDs) (ibid.: 221). These ideas are somewhat contrary to the approach in Chapter 2 of the CLEP report, which suggests that street traders' rights can best be achieved in 'delineated hawker zones' (ibid.: 101). Box 2.1, derived from the report, is innovative in its combination of property and business rights for the informal economy.

Box 2.1 Empowering the informal economy through law

1) Legal and bureaucratic operational procedures
- Simplified registration, licensing and permit procedures
- Identification: ID cards for individuals, and business identification
- Legislation/bylaws, allowing street trading in public space

2) Legal frameworks that enshrine economic rights
- Access to finance, materials and markets
- Access to transport and communications
- Access to improved skills, technology and business services
- Access to business incentives, trade deferrals, subsidies etc.

3) Legal property rights
- Private land
- Intellectual property

4) Use rights to public resources and appropriate zoning regulations
- Use rights to urban public land
- Use of common and public resources
- Appropriate zoning regulations e.g. in CBDs, suburban areas etc.

5) Legal standards for what goods are bought and sold
- Appropriate regulations on legal vs. illegal goods and services
- Product and process standards: eg: public health

6) Legal tools governing transactions and contracts
- Legal and enforceable contracts
- Grievance and conflict resolution mechanisms
- Rights to issue shares, advertise, protect brands etc.

7) Legal rights to social and business protection
- Temporary unemployment relief
- Insurance for land, equipment, and place of work etc.
- Bankruptcy and default rules
- Limited liability, asset and capital protection
- Capital withdrawal and transfer rules

8) Legal rights of association and representation
- Membership of mainstream business associations
- Membership in guilds or product associations
- Representations in planning and rule-setting bodies

Critics of de Soto's approach have argued that the formalization requirement excludes those who cannot afford to pay and discriminates against those without occupancy rights (i.e. tenants), and that the economic benefits of secure title have been overstated (Davis, 2006; Gilbert, 2002). A further critique is that de Soto's distinction between formal rights backed by national legislation and 'extra-legal' rights, and his emphasis on individual title, discriminate against those with well-established collective tenure rights.

Otto (2009) argues that de Soto's work was based on mistaken generalizations, and from a socio-legal study of land regularization in ten countries he argues that de Soto's advocacy of formalization rests on five mistaken assumptions:

- *On law* – that legal systems support the poor when instead they promote neoliberal ideologies of market-led growth;
- *On politics* – that political elites will relinquish control over valuable land, and that individuals are not subject to patronage networks, neither of which hold true;
- *On administration* – that officials are pro-poor and not influenced by ethnic loyalty, bribery or political pressures to promote development, when in practice local governments worldwide are implicated in land grabbing and eviction;
- *On economics* – that legalized plots will remain in the ownership of the poor and that banks will lend on small plots of land, when evidence suggests the contrary;
- *On social norms* – that customary rules can be absorbed into state law which overlooks long traditions of collective rights, and that legal rules can be fixed, precise and neutral, which is rare.

Ibid.

Banik (2009) also argues that while the CLEP report put legal empowerment on the global map, and moved beyond the narrow focus on property formalization, the approach has many shortcomings. For example, the basis on which the four pillars were identified is not clear, nor the reason why other legal issues important for poverty reduction such as education, safety, or women's rights were left out. He suggests that its recommendations remain top-down, state-centred and orthdox (ibid.: 129). Instead, Banik argues, legal empowerment should be seen as a global social contract involving a 'bundle of rights' that provide the poor with the legal and institutional tools to support development and empowerment (Banik, 2008).

In academic spheres, the implication of legal empowerment for the informal economy has been explored by a number of authors. Faundez (2009) argues that the CLEP report's recommendations are unrealistic, and that legal reform needs to be set within a strategy addressing political, social and economic conditions. Banik (2011: 4) further argues the need for collective organization and political mobilization in parallel with legal reform.

Interesting work on the ground is also emerging. For example, in Tanzania, the MKURABITA programme (*Mpango wa Kurasimisha Rasilimali na Biashara za*

Wanyonge Tanzania) (Programme to Formalize the Property and Business of the Poor in Tanzania) worked with de Soto's Institute of Liberty and Democracy (ILD) to address informal-sector assets in Tanzania, and improve tax collection. The four-stage process of Diagnosis, Reform Design, Implementation and Capital Formation, and Governance commenced in early 2005, led by a team from ILD. The Diagnosis, carried out in January–October 2005, identified the scale of informality in the country, estimating that 98% of businesses operated extra-legally with a value of US$ 29.3billion (ILD, 2005) and 50% did not have a licence (MKURABITA, 2007). Reforms have continued through the Business Environment Strengthening programme for Tanzania (BEST). The initial implementation looked at property formalization in 33 district councils and six urban areas, and formalization of 2,000 businesses in Mchikichini and Gerezani; MKURABITA also proposed a raft of small-scale legislative changes to business regulation (key informant interviews).

Thus legal empowerment rests largely on the proposition that poverty reduction can be addressed through reform of existing legal frameworks across three strands: reforming national and local legislation to be more accessible to the poor; addressing the challenges of informality; and formalizing property and business rights. For street traders, there are many gaps in the debates – for example, the assumption that rights should be individual, that collective rights play a muted role and that legal reform will be underpinned by benign and transparent state action.

Right to the city

A third, more radical vision of urban inclusion is found in the *right to the city* debates in Latin America where the claim for the city has underpinned constitutional and legislative change – in Brazil, Ecuador, Colombia and elsewhere. The focus until recently has been on the social function of property for shelter rather than the informal economy, although that is beginning to change.

The term was first coined by French sociologist and philosopher Henri Lefebvre in his 1967 publication, *Le droit à la ville* (*The Right to the City*) which became a *cause célèbre* in the student uprisings and national strikes of May 1968 in France in which around 10 million workers, almost two-thirds of the country's workforce, took part. Lefebvre's concept created a radical paradigm that challenged the social and political structures of the capitalist world (Mitchell, 2003). He argued that the *use value* of cities as centres of cultural and political life were being undermined by processes of industrialization and commercialization, commodifying urban assets for their *exchange value*. For Lefebvre:

> The right to the city manifests itself as a superior form of rights: right to freedom, to individualization in socialization, to habit and to inhabit. The right to the *oeuvre*, to participation and *appropriation* (clearly distinct from the right to property), are implied in the right to the city.
>
> Lefebvre, 1968, in Kofman and Lebas, 1996: 174,
> emphasis in original

Fully realized, Lefebvre's right to the city necessitates a profound reorganization of current social relations (Purcell, 2003; Brown and Kristiansen, 2008). Citizenship is broadly defined to include *all urban inhabitants*, and the status of citizenship is earned by living in the city. Of importance is Lefebvre's emphasis on the right to *the city as a whole*, rather than to specific rights in cities (Brenner, 2001, 2012; Brown, 2013).

More recent academic debates on the right to the city further challenge the nature of capitalism. David Harvey sees the drive for capital accumulation as the central motif in the narrative of historical-geographical transformation of the western world (Harvey, 1973), with capital accumulation contributing to evolving spatial structures, the rise and fall of major cities and the gentrification and commodification of urban space (Castree, 2007). Capitalist investments are continually recycled to increase accumulation, from industry to finance and then real estate, impelling the politics of urban renewal through which property owners, mayors and bureaucrats collude to condemn large swathes of inner-city slums to wrest profit and glory from wholesale demolition (Zukin, 2006).

The process of urban development and production of space is thus a social, economic and political process that reinforces the self-interest of political and social elites, resulting in cities that are increasingly divided, privatized and conflict prone (Harvey, 2008; Marcuse, 2012). Harvey argues that true social change can only be achieved through the constant struggle if transformative politics and grassroots activism constantly mingle with local mobilizations of rebel cities (Harvey, 2003, 2012). Thus, the right to the city is:

> far more than the individual liberty to access urban resources: it is a right to change ourselves by changing the city … The freedom to make and remake our cities is … one of the most precious yet most neglected of our human rights.
>
> Harvey, 2008: 23

The nature of citizenship is a core issue in right to the city debates. Lefebvre's idea is that citizenship is held by all who inhabit the city, not just urban inhabitants, and that to frame citizenship in formal and territorial terms fails to recognize the city as a political community and the social relations of power (Dikeç and Gilbert, 2002). Citizenship confers on individuals certain rights and obligations, including the right to have a voice in the exercise of state power (Brown *et al.*, 2010; Purcell, 2003). The obligations include payment of taxes and acting according to community-defined norms; national voting is an institutionalized requirement but not a mechanism for progressive social change.

This is important for street traders, who are often transient populations, as in China where many traders do not have urban residency status *(hukou)* (see Chapter 10). However, applying Lefebvre's notion poses considerable challenges – for example, should citizenship include only those who pay taxes? Migration – whether forced, economic or environmental; internal or international – is disproportionately focused on cities, and recent figures suggest that, in 2005, 191 million people, or 3% of the world's population, lived outside their country of birth including 17 million in Africa and 58 million in Asia (GMG, 2010: 2). Thus the concept of residency as a basis for citizenship widens the definition of who has a right to the city, but it still omits key constituencies of the urban poor, such as workers in the informal economy, or undocumented migrants. Opinions differ also on the extent to which the right to the city should be enshrined in law; Fernandes (2007), for example, suggests that legal underpinning is essential to its effective operation.

From an urban perspective, Parnell and Pietersee (2010) are among the few that see city governments as important in the adoption of a rights-based agenda; they identify four tiers of citizens' rights: individual rights of voting, freedom and health etc.; collective rights to basic services, including shelter and water; city-scale entitlements such as safety and social amenities; and freedom from human-induced threats such as economic volatility or climate change, but as yet only the first two tiers are widely recognized within current human rights regimes.

'Fuzzy' city governance

City governments are at the front line in managing the urban informal economy. Often with legal powers to protect public order, manage streets and promote urban planning, the actions of local government have a significant effect on poor city dwellers. Yet, modern local government in the developing world faces many challenges, as local administrations remain under-funded and under-staffed, and have little respect in much of the global south.

The concept of *governance* refers to the mechanisms, processes, relationships and institutions through which citizens and groups articulate their interests, both formal and informal, the ways in which decisions over urban management and development take place. Local government law is an important dimension affecting the informal economy as it determines the powers, structure and financing of local government, whether city governments are elected or appointed, boundary specifications and services, taxation and powers to regulate land use. Policy implementation is often opaque, depending on fiefdoms and coalitions of power between politicians and officials. Stone (2004) argues that lower-status groups are weakly positioned to access 'coalitions of power', but social problems can only be addressed if these groups are active partners in change.

Formal systems of local government draw on a raft of legislation that has many plural influences. Constitutional recognition is important to ensure the viability of local governments. The legal framework for local government usually determines the structures and powers of local administrations; leadership and decision-making processes; revenue and financing; and accountability and performance management. However, many other laws and regulations define the remit of local councils covering issues of policing, highways, property management, market regulations, slaughter houses, waste management, or environmental protection.

Many local governments also have a legacy of bylaws and regulations, from multilateral agencies pursuing ideals of 'good governance', 'capacity building' and 'local economic development', of colonial instruments of control. Decentralization has been a major thrust of local government reform, based on the principle that local-level decisions and local administrations will be more responsive to people's needs and more accountable than central provision, but many local authorities lack skills and awareness to support and manage informal economies (Ribot, 2002). As Devas *et al.* (2004: 194–5) conclude, in order to address poverty and exclusion city governments must adopt democratic and participatory processes that move beyond clientelist favours to the delivery of services as a matter of routine and right. Interventions need to value the delicate social and economic relationships that allow the poor to survive; 'bad governance' can destroy livelihoods and jobs of the poor.

Latterly, international agendas have also influenced local government law. Although, in international law, cities have no legal identity, Frug and Barron (2006) argue that international trade agreements, international development policy and human rights investigations are all influencing the way that cities operate. The World Bank and others present cities as neutral agents striving for uncontroversial goals, but international local government law appears to be promoting a particular conception of the 'private city' that sees city power as a way of promoting private wealth in a community of money-makers – interestingly a similar perspective to Harvey from a very different starting point.

Local government financing is a major concern, including the relative importance of central–local transfers and powers to raise local revenue – for example, through property tax, licensing, fees, beer tax etc. While detractors argue that informal economy businesses evade taxes, in practice, informal businesses often pay considerable amounts in taxes, licences or daily fees, including bribes and protection money. The problem is for local governments to eliminate payments to non-state or illicit actors by providing effective legal protection in exchange for payment of taxes and fees, and to provide identifiable benefits in return.

Unofficial pluralism

In additional pluralism *within* state law and bylaws, several *informal processes* affect street trading. First 'fuzzy governance' and *unpredictable implementation* leads to considerable variation in the implementation of statutes and bylaws. The two most important agencies in the day-to-day management of street trade are the city council and police, whose officials are frequently under-paid and lack access to training and equipment. In many smaller and second-tier council offices and even some major cities, computers are still a rarity, access to transport is limited and people may work in one job for many years without prospect of promotion. Where street trading is not high on the political agenda, day-to-day management falls to low-paid officials whose management style may be characterized by low-level harassment, or the collection of 'inducements' on the assumption that street traders are ignorant of the law or powerless to protest. This provides opportunity for small gains of power – for example, field studies in Togo in 2009 found that market managers in the autonomous markets' agency EPAM,[1] in charge of Lomé's Grand Marché, had extended their authority to the surrounding streets where, although beyond EPAM's jurisdiction, traders could not resist the fees and trading controls (Lyons and Brown, 2010).

Second, *political pressures* create dynamics of change in which legal instruments serve as a convenient tool to consolidate power. Street traders often complain that the only time they see politicians is in the run-up to municipal or national elections (interviews). For example, in the 2000 election in Sénégal, street traders were an important voting bloc in returning President Aboulaye Wade to power. After an election or new appointment the pressures change, as new post-holders seek results-oriented action and favourable media headlines to demonstrate their effectiveness. The outcome is often an enforcement of bylaws and a campaign to 'clean up the streets'. For example, in Tanzania in March 2006, shortly after President Kikwete came to power at the end of 2005, an order to remove street traders in all major towns was issued, eventually affecting perhaps a million traders (Lyons and Msoka, 2009). Nigeria's Kick Against Indiscipline operation, set up within the Lagos State Ministry of Environment, focuses on the eradication of street trading (Omoegun, 2015).

International political agendas also have an impact on street traders' operations. Setšabi (2006) scanned newspaper reports in Maseru between 1993 and 2003 to demonstrate how street trader evictions often took place before the visit of foreign dignitaries. Mega-sporting events provide a more extreme version of legal enforcement for political ends as land is cleared along the route from the airport or near major sporting facilities (Duminy and Luckett, 2012). Street traders in Delhi said that around 50 trading sites had been cleared for the 2010 Commonwealth Games and only about half had returned (fieldwork interviews, 2011). The neoliberal agenda promoted by multilateral agencies can have a significant impact. For example, in Tanzania government ministries were tasked with improving the country's rankings in the World Bank's *Doing Business* report, an economic ranking based on indicators of 'ease of doing business', although many of the reforms undermined the operation of micro-businesses such as street traders (Lyons *et al.*, 2012).

Third, *street-level dynamics* often transcend local bylaws, with considerable variation in the effectiveness of street trader organizations. The associational landscape of street and market trade is extremely varied and has been covered at length elsewhere (e.g. Brown *et al.*, 2010; Lindell, 2010). Street trader associations vary in their aims, capacity and effectiveness, and those which focus on solidarity and welfare may have less purchase than those which adopt an advocacy or campaigning role (Brown *et al.*, 2010). Some associations may be formal or semi-formal, such as the Mchikichini Marketing Cooperative (MCHIMACA) in Dar es Salaam – a registered umbrella association for six sector associations for traders in new and second-hand household goods that works with a microfinance association, and managed the market of 10,000 traders under contract to the municipal council (interviews). Others have traditional roots, such as Ghana's product associations, an off-shoot of powerful pre-colonial guilds chaired by a hereditary queen or king, which are closely involved in market management (King, 2006). In Sénégal the powerful Mouride religious brotherhoods and smaller Layennes and Tijanes dominated trading networks in Dakar's central Marché Sandaga, and the brotherhood supported supply chain networks and guaranteed access to space (Brown *et al.*, 2010), before several disastrous fires destroyed the market in 2013.

However, many traders shun associations, which they see as expensive or ineffective, yet conform to a plethora of unwritten customs and procedures. Access to trading space is often controlled by gatekeepers, perhaps community elders, adjacent landowners, or long-established traders, who sometimes have benign relationships with traders. Kinship networks are also extremely important. In Ahmedabad, many areas are controlled by an *aagyawan*, a local protector who collects 'rent', manages bribe payments, negotiates with city officials and the police to protect his group and may keep written records of traders on his patch. Sometimes such protectors work in collusion with city officials; for example, again in Ahmedabad, traders reported an increase in *hafta* payments (backhanders) when a new police superintendent was appointed, as lowly police sergeants were pressured into paying more so that senior staff could also return favours.

The key issue for the informal economy is *how local governments harness formal urban law* to manage large numbers of independent operators in the informal economy, and what *informal processes provide opportunities for cooperative management*. It is difficult for poorly resourced local governments to process large numbers of small transactions – for example, collecting fees from street traders – and for the disparate enterprises in the informal economy to combine interests, particularly when many of those transactions are in cash. The bargaining power of traders is often weak and, in general, they lose out from major urban development and the privatization of public services. For example, waste-pickers lose out when waste collection is privatized, and street traders are often dispossessed in historic area improvements (Crossa, 2009) or by the introduction of new urban freeways and rapid transit routes. Nevertheless, the practices of spatial claim (Mackie *et al.*, 2014) and collaborative management by street traders can provide important pointers for the future.

Conclusion

It is clear that legislative frameworks affecting street trading are complex and contradictory, but the understanding of legal pluralism discussed above suggests that the regulation of street trading operates across five domains of complexity: the written statues and bylaws of state law; international commercial and business law; local implementation of the law; local and international politics; and informal norms of conduct that govern street trade.

The legal pluralism debate suggests that social ordering influences street trade in a myriad of ways and in fluid urban societies will be continually adapting, highlighting the mobile and contingent nature of the law. Plural legal influences affecting the working poor occur in different ways – partly from the numerous influences within formal state law, influenced by national and international agendas, moderated by local practice, but also by the multiplicity of informal practices and internal regulatory process of the urban informal economy. Collective systems of order and justice and their relations with local government administrations have unexplored potential for strengthening the livelihoods of the poor. The challenge is to harness this informal dynamic to support urban livelihoods; although the theory is interesting, its translation into practice is less well addressed.

The foundation of the legal empowerment agenda is flawed on a number of counts. In addition to the mistaken assumptions that Otto identifies, its fundamental tenets are that formalization is the only solution; poverty reduction is the responsibility of the individual; and tinkering with law at the margins will enable individuals to accumulate capital and assets and beat a path out of poverty. While de Soto's work in Tanzania has produced some interesting recommendations for legal reform, his analysis of *L'économie informelle* in Tunisia is less robust (de Soto, 2012). The core problem is that poverty is the result of a dynamic between the individual, the state and wider market-led processes of change, which the legal empowerment agenda does not fully address. Academic debate has taken these ideas forward, but as yet to little practical effect.

In contrast, the right to the city is a powerful concept that tackles the drivers of poverty and exclusion in market-led systems of exchange by achieving collective empowerment and an effective dialogue between citizen and state. Its implementation is difficult but it has operated effectively in Brazil and Ecuador (see Chapters 4 and 5) – the problem is that ideals which have widespread acceptance in Latin America have little traction in other parts of the world – for example, the right to the city does not translate easily into Arabic or Chinese.

The philosophy proposed here is that, without collective empowerment – which may or may not be formally expressed, for example, through registered cooperatives or microfinance organizations – the poor will not achieve a critical mass to effect change. Collective action may take many forms – families and kinship groups, savings groups, religious organizations or market associations – and urban areas should embrace such plurality, although checks and balances are required. Conflicting and plural influences within legal and regulatory systems need to be addressed, but legal reform must reflect the informal systems and practice that underpin informal economy livelihoods. Legislative change is important, particularly enshrining the right to a decent livelihood in constitutions, removing conflicting and multiple sources of illegality, and giving legitimacy to informal economy workers, but it is not sufficient to effect change. Change will not take place without strengthening the recognition and capacity of local governments and recognizing their crucial role in supporting the livelihoods of the poor.

Note

1 Etablissement Public Local Autonome pour l'Exploitation des Marchés de Lomé.

References

Banik, D. (2008) Rights, legal empowerment and poverty: an overview of the issues, in Banik, D. (ed.), *Rights and Legal Empowerment in Eradicating Poverty*

Banik, D. (2009) Legal empowerment as a conceptual and operational t, Farnham and Burlington, VT: Ashgate.ool in poverty eradication, *Hague Journal of the Rule of Law*, 1(1): 117–31.

Banik, D. (ed.) (2011) *The Legal Empowerment Agenda: Poverty, Labour and the Informal Economy in Africa*, Farnham and Burlington, VT: Ashgate.

Benda-Beckmann, F. von (2002) Who's afraid of legal pluralism?, *Journal of Legal Pluralism*, 47(1): 37–83.

Brenner, N. (2001) State theory in the political conjuncture: Henri Lefebvre's 'Comments on a new state form', *Antipode*, 33(5): 783–808.

Brenner, N. (2012) What is critical urban theory?, in Brenner, N., Marcuse, P. and Meyer, M. (eds), *Cities for People, Not for Profit: Critical Urban Theory and the Right to the City*, London and New York: Routledge.

Brown, A. (2006), *Contested Space: Street Trading, Public Space and Livelihoods in Developing Cities*, Rugby: ITDG.

Brown, A. (2013) 'Right to the City': the road to Rio, *International Journal of Urban and Regional Research*, 37(3): 957–71.

Brown, A. and Kristiansen, A. (2008) *Urban Policies and the Right to the City: Rights, Responsibilities and Citizenship*, UNESCO, MOST (Management of Social Transformations) Programme: Paris, http://unesdoc.unesco.org/images/0017/001780/178090e.pdf, accessed August 2010.

Brown, A., Lyons, M. and Dankoco, I. (2010) Street traders and the emerging spaces for urban voice and citizenship in African cities, *Urban Studies*, 47(3): 666–83.

Castree, N. (2007) David Harvey: Marxism, capitalism and the geographical imagination, *New Political Economy*, 12(1): 97–115.

Croce, M. (2012) All law is plural: legal pluralism and the distinctiveness of law, *Journal of Legal Pluralism*, 65: 1–30.

Crossa, V. (2009) Resisting the entrepreneurial city: street vendors' struggle in Mexico City's historic center, *International Journal of Urban and Regional Research*, 33(1): 43–63.

Davis, M. (2006) *Planet of Slums*, London: Verso.

de Soto, H. (1989) *The Other Path: The Invisible Revolution in the Third World*, New York: Perennial.

de Soto, H. (2000) *The Mystery of Capital: Why Capitalism Triumphs in the West and Fails Everywhere Else*, New York: Basic.

de Soto, H. (2012) *L'économie informelle. Comment y remédier? Un opportunité pour la Tunisie*, Tunis: UTICA, Lima: ILD.

Devas, N. with Amis, P., Beall, J., Grant, U., Mitlin, D., Nunan, F. and Rakodi, C. (2004) *Urban Governance, Voice and Poverty in the Developing World*, London and Sterling, VA: Earthscan.

Dikeç, M. and Gilbert, L. (2002) Right to the city: homage or a new social ethics?, *Capitalism Nature Socialism*, 14(2): 58–74.

Duminy, J. and Luckett, T. (2012), Literature survey: mega-events and the working poor, with a special reference to the 2010 FIFA World Cup, African Centre for Cities, University of Cape Town, http://www.inclusivecities.org/wp-content/uploads/2012/07/Duminy_Luckett_Mega_events_Working_Poor.pdf, accessed July 2015.

Faundez, J. (2009) Empowering workers in the informal economy, *Hague Journal of the Rule of Law*, 1(1): 156–72.

Fernandes, E. (2007) Constructing the 'right to the city' in Brazil, *Social Legal Studies*, 16(2): 201–19.

Frug, G. and Barron, D. (2006) International local government law, *The Urban Lawyer*, 38(1): 1–62.

Gilbert, A. (2002). On the mystery of capital and the myths of Hernando de Soto: what difference does legal title make?, *International Development Planning Review*, 24(1): 1–19.

GMG (2010) *International Migration and Human Rights: Challenges and Opportunities on the Threshold of the 60th Anniversary of the Universal Declaration of Human Rights*, New York: Global Migration Group (GMG), UNDP.

Griffiths, A. (2011) Pursuing legal pluralism: the power of paradigms in a global world, *Journal of Legal Pluralism and Unofficial Law*, 64: 173–202.

Harvey, D. (1973) *Social Justice and the City*, London: Edward Arnold.

Harvey, D. (2003) The right to the city, *International Journal of Urban and Regional Research*, 27(4): 939–41.

Harvey, D. (2008) The right to the city, *New Left Review*, 53, September/October: 23–40.

Harvey, D. (2012) *Rebel Cities: From the Right to the City to the Urban Revolution*, London and New York: Verso.

Herrera, J., Kuépié, M., Nordman, C.J., Oudin, X. and Roubaud, F. (2012) Informal sector and informal employment: overview of data for 11 cities in 10 developing countries, WIEGO Working Paper No. 9, http://wiego.org/sites/wiego.org/files/publications/files/Herrera_WIEGO_WP9.pdf, accessed August 2015.

ILD (2005) *Volume I: Executive Summary, The Diagnosis*, http://www.mkurabita.go.tz/userfiles/2007-9-4-5-28-0_executive%20summary.pdf, accessed July 2015.

ILO (2013a) *Women and Men in the Informal Economy: A Statistical Picture*, 2nd edn, Geneva: International Labour Office (ILO), http://www.ilo.org/stat/Publications/WCMS_234413/lang--en/index.htm, accessed July 2015.

ILO (2013b) *Transitioning from the Informal to Formal Economy*, Geneva: ILO, http://www.ilo.org/wcmsp5/groups/public/---ed_norm/---relconf/documents/meetingdocument/wcms_218128.pdf, accessed July 2015.

King, R. (2006) Fulcrum of the urban economy: governance and street livelihoods in Kumasi, Ghana, in Brown, A. (ed.), *Contested Space: Street Trading, Public Space and Livelihoods in Developing Cities*, Rugby: ITDG.

Kofman, E. and Lebas, E. (eds and trans) (1996) *Writings on Cities*, Oxford: Blackwell.

Lefebvre, H. (1968) *Le droit à la ville*, in Kofman, E. and Lebas, E. (eds and trans), *Writings on Cities*, Oxford: Blackwell.

Lindell, I. (ed.) (2010) *Africa's Informal Workers*, London: Zed.

Lyons, M. and Brown, A. (2010) Has mercantilism reduced urban poverty in SSA? Perception of boom, bust and the China–Africa trade in Lomé and Bamako, *World Development*, 38(5): 771–82.

Lyons, M. and Msoka, C. (2009) The World Bank and the African street: (how) have the doing business reforms affected Tanzania's micro-vendors?, *Urban Studies*, 47(5): 1079–109.

Lyons, M., Brown, A. and Msoka, C. (2012) (Why) have pro-poor policies failed Africa's working poor?, *Journal of International Development*, 24(8): 1008–29.

Mackie, P., Bromley, R. and Brown, A. (2014) Informal traders and the battlegrounds of revanchism in Latin America, *International Journal of Urban and Regional Research*, 38(5): 1884–903.

Marcuse, P. (2012) Whose right(s) to what city?, in Brenner, N., Marcuse, P. and Meyer, M. (eds), *Cities for People, Not for Profit: Critical Urban Theory and the Right to the City*, London and New York: Routledge.

McAuslan, P. (2005) Legal pluralism as a policy option: is it desirable, is it doable?, in UNDP–International Land Coalition Workshop, *Land Rights for African Development: From Knowledge to Action Nairobi*, CAPRi Policy Briefs, https://commdev.org/userfiles/capri_brief_land_rights.pdf, accessed June 2015.

Merry, S. (1988) Legal pluralism, *Law and Society Review*, 22(5): 869–96.

Mitchell, D. (2003) *The Right to the City: Social Justice and the Fight for Public Space*, New York and London: Guilford Press.

MKURABITA (2007) Baseline study of informal businesses and properties in Dar es Salaam, http://www.mkurabita.go.tz/resource_centre/documents.php, accessed June 2013.

Moore, S. (1973) Law and social change: the semi-autonomous social field as an appropriate subject of study, *Law and Society Review*, 7(1): 719–46.

Omoegun, A. (2015) Street trader displacements and the relevance of the Right to the City concept in a rapidly urbanising African city: Lagos, Nigeria, PhD thesis, Cardiff University, http://orca.cf.ac.uk/72513/1/Ademola%20Omoegun%20Thesis.pdf, accessed July 2015.

Otto, J.M. (2009) Rule of law promotion, land tenure and poverty alleviation: questioning the assumptions of Hernando de Soto, *Hague Journal of the Rule of Law*, 1(1): 173–95.

Parnell, S. and Pieterse, E. (2010) The 'right to the city': institutional imperatives of a developmental state, *International Journal of Urban and Regional Research*, 34(1): 146–62.

Purcell, M. (2003) Citizenship and the right to the global city: reimagining the capitalist world order, *International Journal of Urban and Regional Research*, 27(3): 564–90.

Ribot, J. (2002) African decentralization: actors powers and accountability, United Nations Research Institute for Social Development (UNRISD) Programme on Democracy, Governance and Human Rights, Paper No: 8, Geneva: UNRISD.

Santos, B. de Souza (1987) Law: a map of misreading. Towards a post-modern conception of law, *Journal of Law and Society*, 14(3): 279–302.

Santos, B. de Souza (2006) The heterogeneous state and legal pluralism in Mozambique, *Law and Society Review*, 40(1): 39–75.

Setšabi, S. (2006) Contest and conflict: governance and street livelihoods in Maseru, Lesotho, in Brown, A. (ed.) *Contested Space: Street Trading, Public Space and Livelihoods in Developing Cities*, Rugby: ITDG.

Stone, C.N. (2004) It's more than the economy after all: continuing the debate about urban regimes, *Journal of Urban Affairs*, 26: 1–19.

Snyder, F. (1981) Colonialism and legal form: the creation of 'customary law' in Senegal, *Journal of Legal Pluralism and Unofficial Law*, 19: 49–90.

Tamanaha, B. (1993) The folly of the 'social scientific' concept of legal pluralism, *Journal of Law and Society*, 20(2): 192–217.

UNDP (2008) *Making the Law Work for Everyone: Volume 2, Report of the Commission on Legal Empowerment of the Poor*, New York: UNDP.

Winters, M. and Kulkarni, S. (2014) The World Bank in the post-structural adjustment era, *Handbook of Global Economic Governance: Players, Power and Paradigms*, New York: Routledge.

Zukin, S. (2006) David Harvey on Cities, in Crastree, N. and Gregory, D. (eds), *David Harvey: A Critical Reader*, Oxford: Blackwell.

3 Rights-based approaches and social injustice

A critique

Beth Watts and Suzanne Fitzpatrick

Summary

Chapter 3 distinguishes between different kinds of rights: *natural rights* and *human rights* in the 'universal realm' and *programmatic rights* and *positive legal rights* in the 'national realm'. It thus provides a context for understanding rights-claims in relation to the informal economy. The chapter further considers unintended consequences that may result from an uncritical pursuit of *absolute* rights. It concludes by emphasising the importance of maintaining a critical perspective on rights discourses as a means of addressing social injustice, on both philosophical and pragmatic grounds.

Introduction

'Rights-talk' carries with it considerable intuitive force, with 'rights-based approaches' promising radical solutions to complex issues of poverty, marginalisation and social exclusion. Such approaches hold the prospect of challenging existing distributions of economic, political and social power, thereby empowering disadvantaged and marginalised groups and overcoming the stigma of dependence on charitable or discretionary welfare assistance. However, beneath this ostensible appeal lie fundamental conceptual and empirical questions: what precisely do we mean by rights-based approaches, and do they deliver the things we expect them to in practice?

At its core, the idea of 'rights' conveys that the 'rights-holder' or 'possessor' has some form of entitlement. However, 'rights' can be moral or legal, abstract or specific, enforceable or unenforceable, and national or international. They can be seen as absolute side-constraints on the pursuit of other objectives (such as economic growth) or as social goals that may conflict and have to be traded-off against each other (Fitzpatrick and Watts, 2010; Bengtsson *et al.*, 2012; Attoh, 2011). Conceptually, 'rights' are something of 'a black box' (Attoh, 2011: 669). Starting from an acknowledgement of this complexity, this chapter critically examines the efficacy of 'rights-based approaches' in tackling social injustice, exploring their philosophical foundations as well as their practical applications.

We begin by considering the relevance of the centuries-old debate about the existence or otherwise of the *natural rights* of human beings, before moving on to consider *human rights*, in many ways the modern cousin of natural rights. Practical and philosophical objections to both of these sorts of *moral rights* in attempting to address substantive social needs – and responses to such objections – are considered. The focus then shifts from the international or 'universal' realm, to rights discourses in the national, domestic realm. Here it is necessary to consider *citizenship rights*, which are *programmatic* in nature, as well as *positive legal rights* which are individually enforceable through domestic courts. Again the merits and demerits of these forms of rights are considered with respect to pursuing social justice for disadvantaged groups.

Two key assumptions underpin this chapter. First, that it is possible to object to natural and/or human rights and still be in favour of clearly delimited 'positive' legal rights, and vice versa; and second, that human rights offer only one among several (progressive) normative frameworks through which to consider issues of social exclusion and disadvantage. As will become clear, the critique of right-based approaches offered here is most relevant in contexts of *relative scarcity* such as that which pertains in advanced economies which have developed at least minimalist welfare states. We appreciate that some of these arguments may play out differently in contexts of *absolute scarcity* and in those parts of the world which lack functioning constitutional democracies.

The universal realm: natural and human rights

This section focuses on universal moral rights, as understood from the perspectives of natural law and human rights. It considers first, how rights are understood from these perspectives, before presenting a series of critiques of these ways of thinking about rights.

Natural law and natural rights

Natural or *doctrinal* rights refer to a set of universal, inalienable rights held by all human beings (Norman, 1998; Dean, 2002). This conception of rights began to emerge as part of the western Enlightenment during the seventeenth and eighteenth centuries, building on the ideas of classical philosophers like John Locke (1690). Bills of Rights in England (1689), America (1789) and France (1789) reflected an understanding that individuals were the bearers of rights. This liberal tradition conceives of rights as fundamental, bestowed by God, or another divine source, or by some understanding of the nature of humanity. Natural rights have largely been concerned with people's civil and political rights, rather than social rights to substantive welfare entitlements. Nevertheless, as early as 1791 Thomas Paine argued that 'poor relief' under the 'Poor Law' ought to be replaced with a 'right to relief' (Dean, 2010).

The jurisprudential roots of natural rights are to be found in the natural law tradition, which holds that 'what naturally is, ought to be' (Finch, 1979: 29).

In other words, the 'law of nature' should be used as a standard against which one can measure the validity or rightness of man-made law. Over the centuries, legal theorists have sought to derive this 'law of nature' variously from 'universal nature', 'divine nature' and 'human nature' itself. A key philosophical reference point for natural rights lies in the Immanuel Kant-inspired (Kantian) school of moral philosophy. According to this 'deontological' style of ethics, an action is deemed morally right or wrong on the basis of the natural or 'universalisable' duties people owe to each other, in line with the 'categorical imperative' to always do one's duty regardless of the consequences. A rights-based approach is most often interpreted as deontological because rights can be seen as rules, side-constraints or 'trumps' (Dworkin, 1977) that (ethically) limit the actions that can be taken against individuals in order to pursue collective goals.

Deontological ethics are normally contrasted with consequentialist moral theories. Consequentialism dictates that morally 'good' actions are those which tend to bring about 'valuable states of affairs' (Williams, 1995). The most influential strand of consequentialist ethics – utilitarianism – supports actions which maximise the sum total of societal 'welfare', popularly referred to as the 'greatest happiness of the greatest number'. Utilitarian thinkers, especially Jeremy Bentham (1789), are closely associated with the 'legal positivist' strand within jurisprudential scholarship, which has highlighted the reactionary implications of the 'absolutist' natural law doctrine, and the way in which its speculative character leaves it open to abuse: 'natural law is at the disposal of everyone. The ideology does not exist that cannot be defended by an appeal to natural law' (Ross, 1974: 261).

However, utilitarianism is also open to some obvious objections, not least its apparent disregard for the distribution of well-being, and for failing to respect people (in Kant's famous formulation) as ends and not means. These weaknesses go a long way to explaining the continuing appeal of deontological – and specifically human rights-based – philosophical approaches in the modern era.

Human rights

Human rights are, in many ways, the modern successor to traditional notions of natural law and natural rights (Turner, 1993). Human rights find expression in international instruments, many of which encompass social and economic rights, as well as the civil and political rights associated with natural rights. For example, Article 25 of the United Nations Universal Declaration of Human Rights (1948) asserts: 'Everyone has the right to a standard of living adequate for the health and wellbeing of himself and his family, including food, clothing, housing and medical care and necessary social services.'

While this resolution is not formally binding, it is considered a key part of international customary law and has provided the principal foundation for subsequent debate on universal human rights. Other international instruments do impose obligations on ratifying states, binding in international law. At the international level, the UN Covenant on Economic, Social and Cultural Rights (1966) obliges signatory states to take steps, to the maximum of available resources, to achieve progressively the full realisation of the rights stipulated in the Covenant. At the European level, the key human rights instruments are the Charter of Fundamental Rights (European Union) (2000) and the European Social Charter (Council of Europe) (1961, revised in 1996), the latter of which now incorporates a Collective Complaints mechanism that can be used to expose violations of the Charter by ratifying states.

In relation to workers in the informal economy, it has sometimes been argued in the courts that the right to an adequate standard of living (and the 'right to life' also enshrined in various international human rights instruments) is dependent on a 'right to a livelihood', albeit with mixed success (see Chapters 8 and 9).

According to human rights advocates, every human being *ought* to have access to the rights specified in these international instruments, and nation states – as well as relevant international bodies – *ought* to ensure their delivery. Human rights then – like natural rights – can be understood as *moral* statements about human beings (Fortman, 2006; Fitzpatrick and Watts, 2010). They are 'moral claims on the behaviour of individual and collective agents, and on the design of social arrangements' (UNDP, 2000: 25). They do not necessarily imply duties on the part of specific, legally defined agents to undertake tightly defined acts ('perfect duties' in Kantian terminology), but rather a 'general and non-compulsive duty' (ibid.: 24) on all of humankind ('imperfect duties'). Thus the mark of human rights is that they are 'held by each individual against the whole world' (Waldron, 1993: 23). Neglect of the demands they imply therefore 'involves a serious moral – or political – failure' (UNDP, 2000: 24). This kind of perspective underpins Nussbaum's assertion that there is a 'tragic aspect to a rights violation' (2011: 34).

'Human rights-talk' thus carries with it considerable ethical and intuitive force, and has become an increasingly dominant form of moral discourse in both the developed and developing worlds. But beneath its ostensible appeal lie fundamental conceptual and empirical questions. Four key critiques of human rights are considered here.

First, intrinsic to the notion of human rights is the idea that they are self-evident, inalienable and non-negotiable: 'absolute', in other words. Speaking in these terms, however, could be regarded as a mere rhetorical device, an act of politics, that aims to shut down debate by investing one's own particular policy priorities with a 'protected' status. After all, as Dworkin (1977) insisted, 'rights are trumps'. To borrow Arendt's terminology, the notion of rights as self-evident and non-negotiable 'reduces politics to nature' (1963: 99) in a way that is, at the very least, in need of careful justification and, at worst, highly problematic.

Acknowledging that appeals to divine or natural law theories are deeply problematic (see above), human rights advocates tend to defend the protected status of the rights they espouse on the basis that they are socially constructed and intersubjective, rooted in a broad normative consensus about the things that all human beings are morally entitled to in order to attain a basic standard of living and to participate in society (Dean, 2010). But the idea that such a consensus exists at a global level is, at the very least, arguable (Finch, 1979; Miller, 1999), reflecting the profound challenge involved in justifying universal moral norms in a 'post-metaphysical age' (Lukes, 2008: 117).

Nonetheless, Lukes (2008) sets out to defend just such a contemporary objective morality, asking whether 'one [can] identify components of wellbeing that are present within any life that goes well rather than badly: conditions of human flourishing?' (129). For the answer, he looks to the Aristotelian-inspired 'capabilities approach' (Sen, 1992) which seeks to minimise inequalities in the 'positive freedom' that people enjoy to achieve 'valuable functionings' in key aspects of their lives. In particular, Lukes draws on Nussbaum's (2000) development of the approach and list of ten 'central human capabilities', which includes the following: life; bodily health; bodily integrity; senses, imagination and thought; emotions; practical reason; affiliation; other species; play; and control over one's environment. Nussbaum claims that these capabilities derive from: 'an intuitively powerful idea of truly human functioning that has roots in many different traditions and is independent of any particular metaphysical or religious view' (ibid.: 101).

Crucially, if these capabilities are accepted as essential human functions, this opens up the possibility of their becoming politicised as human rights and constitutional entitlements. Nussbaum makes this link via the notion that 'the ten central capabilities are fundamental entitlements inherent in the very idea of minimum social justice, or a life worthy of human dignity' (ibid.: 25).

Others have proposed alternative (secular) foundations to human rights. Norman (1998), for example, argues that a derivative concept of rights can be based on the satisfaction of basic human needs, as there are rational and objective ways of determining what these needs are. Similarly, Dean (2008), following Fraser (1989), has argued that through a 'politics of needs interpretation' claims may be translated into assertions of rights. At an even more basic level, Turner (1993) argues that, in the absence of natural law, the philosophical foundations of human rights can most effectively be defended via an appeal to the universal nature of human frailty, particularly the frailty of the body.

These sorts of arguments – based on the common-sense premise that people have a right to what they need – are intuitively appealing. There are philosophical objections, however, to deriving an 'Ought' (a value statement) from an 'Is' (a factual statement) in this way (see Norman, 1998). Moreover, even if it is accepted that statements about human need can provide a 'bridge' from statements of fact to statements of value, it has been argued that needs cannot simply be translated into rights. McLachlan (1998) argues that there are many things people need but that cannot be guaranteed to them as a right. Ignatieff (1984), for instance, argues that love, belonging, dignity and respect are all things that we need which cannot be provided within a formal framework of rights.

The second common critique of human rights concerns their lack of enforceability within current institutional contexts. Scruton (2006), an ardent critic of the human rights discourse, powerfully articulates this objection:

> Rights do not come into existence merely because they are declared. They come into existence because they can be enforced. They can be enforced

only where there is a rule of law ... Outside the nation state those conditions have never arisen in modern times ... When embedded in the law of nation states, therefore, rights become realities; when declared by transnational committees they remain in the realm of dreams – or, if you prefer Bentham's expression 'nonsense on stilts'.

Ibid.: 20–1

Commentators sympathetic to – and indeed advocating – a human rights-based approach also acknowledge this issue. Arendt (1973), writing after two world wars which had killed and displaced millions of people, exposed the limits and 'hopeless idealism' (ibid: 269) of the human rights discourse, and, in particular:

the discrepancy between the efforts of well-meaning idealists who stubbornly insist on regarding as 'inalienable' those human rights, which are enjoyed only by citizens of the most prosperous and civilised countries, and the situation of the rightless themselves.

Ibid.: 279

More recently, Fortman (2006) has identified an 'implementation deficit', describing 'the world of human rights [as] a world of unfulfilled expectations' (37). The response of most human rights advocates to this implementation 'gap' is to seek to narrow it through stronger systems of international governance and accountability, in order to realise enforceable rights beyond the boundaries of the national state (e.g. Kenna, 2005).

However, if such an exercise were to bear fruit, this would bring us to the third key objection to human rights approaches in tackling social injustice: their potentially undemocratic nature (Campbell *et al.*, 2011; Tucker, 2015). This objection is prompted by the observation that the 'rights' expressed in international instruments are, inevitably, broad and abstract in nature rather than detailed, delimited and contextualised. If such abstract rights were *in fact* to be rendered routinely enforceable via courts (international or domestic) this would amount to a major transfer of policy-making power from the political to the legal sphere. The term 'over-socialisation' has been used to describe the situation wherein courts are inappropriately used to decide policy issues (Dean, 2002), and there are legitimate concerns about judges rather than (one would hope, democratically elected) politicians setting broad policy aims and priorities. Particularly with respect to social rights to welfare, the prospect of (unelected) judges determining the allocation of scarce resources in situations where 'hard choices' have to be made between a range of needy and/or deserving cases is not an attractive one (see also King, 2003), at least in contexts where there is a fully functioning democratic government.

Foregrounding these issues of scarcity, conflict and prioritisation brings us to our fourth and final potential objection to human rights approaches: the evident reality that these rights can clash, and that such clashes are likelier the broader, more abstract and more ambitious the rights in question are. Just one example, focusing on the much vaunted 'right to the city', will serve to make this point. As Attoh's (2011) excellent critique demonstrates, the 'right to the city' has, on the one hand, been used to defend those engaged in street activities like begging from exclusionary tactics designed to promote the so-called 'revanchist sanitisation of public space' (Johnsen and Fitzpatrick, 2010). At the very same time, on the other hand, it has been used to argue for the democratisation of decision-making concerning spaces in the city, and thus has embedded within it a democratic and majoritarian logic. But it is more than conceivable (it is, in fact, highly likely) that local residents and business owners would prefer to exclude activities like begging from their city. Viewed in this light, the 'right to the city' is both 'a collective right to manage urban resources ... [and] a right against such management' (Attoh, 2011: 677). How does one resolve such conflicts in a rights-based normative framework founded on absolutism? As Waldron has commented:

> an insistence on absolutism does not make the conflicts go away; it doesn't make situations that appear to call for trade-offs disappear. Those situations [of conflict] are not something that consequentialists and their fellow travellers have *invented* to embarrass moral absolutists.
>
> Waldron, 1993: 32, emphasis in original

Crucially, many of these conflicts are not ones which can be resolved simply by directing more resources at an issue (Waldron, 1993). Rather, they arise from the multiplicity of incommensurate social goals that people may (legitimately) have. As Isaiah Berlin has famously argued (1969/2002):

> the belief that some single formula can in principle be found whereby all the diverse ends of men can be harmoniously realised is demonstrably false. If as I believe the ends of men are many, and not all of them are in principle compatible with each other, the possibility of conflict – and of tragedy – can never wholly be eliminated from human life, either personal or social. The necessity of choosing between absolute claims is then an inescapable characteristic of the human condition.
>
> Ibid.: 214

If such 'value pluralism' is accepted as a guiding normative principle, human rights-based approaches clearly run into trouble in their insistence that such rights are inalienable, indivisible and absolute. Sen (1984) (in opposition to Nussbaum) appears to acknowledge this in his 'broad consequentialist' approach that sees human rights not as absolute side-constraints, but rather as social goals to be pursued alongside other legitimate social objectives (see also Alexander, 2004). However, Waldron's response to the inevitability of rights clashes is to call for (socioeconomic) rights to be 'integrated into a general theory of justice, which will address in a principled way whatever trade-offs and balancing are necessary for their institutionalisation in a world characterised by scarcity and conflict' (1993: 33). While noting that the modern preoccupation with rights is in large part a response to the 'brutal aggregative trade-offs' (ibid.: 27) associated with utilitarianism, Waldron argues that modern theories of social justice (notably John Rawls's (1971/1999) classic account) can provide a helpful 'matrix of compromise' that takes seriously the intrinsic worth of each individual human being. But if irreconcilable clashes between human rights have to be resolved via appeal to such an overarching theory of justice, might intellectual and moral honesty not be better served by simply focusing on these core issues of individual-level distributive justice from the outset?

Yet, in spite of all these weaknesses, human rights discourses retain a key strength, which is especially important in countries where democratic traditions and the protection of minorities remain weak or underdeveloped. 'The articulation of imperfect duties' (UNDP, 2000: 26) encompassed in the human rights framework has been persuasively argued to constitute a 'discursive resource' (Dean, 2008: 9) with 'rhetorical and agitprop merits … when it comes to exposition or to "consciousness raising"' (UNDP, 2000: 24). From this perspective, human rights are 'political instruments to mobilize dissent, protest, opposition and collective action aimed at social and economic reform' (Fortman, 2006: 38, Waldron, 1993). As Isaac (2002) argues, it is irresponsible to simply deconstruct and expose the weaknesses of the human rights discourse without proposing alternative, superior ways of pursuing social justice or, at least, humanitarian goals on a global basis (see also Miller, 1999). Even Ignatieff – a sceptic of the rights discourse in many regards – sees human rights as a valuable 'shared vocabulary' (2000: 349).

So, for all their philosophical and practical limitations, human rights may be considered a 'useful fiction' (Fitzpatrick *et al.*, 2014), justified, perhaps ironically, on the consequentialist basis that in all likelihood they do more good than harm (see also Sen, 1984).

The national realm: social citizenship, programmatic and legal rights

Our discussion thus far has focused on rights at the international level. But there are also relevant rights discourses at the national level: in fact, the concept of 'citizenship rights' predates that of human rights by some considerable margin (Dean, 2010). The classic account of the development of 'social' citizenship rights in the post-war era is by T.H. Marshall (1949), albeit that his 'evolutionary' account (part empirical, part normative) is heavily influenced by the specific UK experience (Turner, 1993). In contemporary debate, social rights – in contradistinction to civil or political rights – have been defined as substantive entitlements to goods or services owed to individuals by the state (Dean, 2002). However, social citizenship rights at the national level can take a variety of forms (Bengtsson, 2001; Dean, 2002). In particular, a distinction must be drawn between 'legal' or 'positive' social rights, on the one hand, and 'programmatic' social rights, on the other.

Programmatic rights

Programmatic rights can be seen as the equivalent of moral or human rights at the national level (Bengtson *et al.*, 2012), insofar as they impose a 'general and non-compulsive duty' on key national actors to pursue certain goals. A programmatic approach to rights 'binds the State and public authorities only to the development and implementation of social policies, rather than to the legal protection of individuals' (Kenna and Uhry, 2006: 1).

Programmatic rights have been argued to be important insofar as they 'express goals which political actors ... agree to pursue' (Mabbett, 2005: 98), and Bengtsson (2001: 255) has described such rights as 'political markers of concern'. Talking about housing specifically, Bengtsson associates programmatic rights with 'universalistic' welfare regimes, as opposed to more legalistic rights, which he associates with selective welfare regimes (see below). Interestingly, he highlights that the former interpretation of rights reflects Marshall's (1949) original (but oft misunderstood) conception of social rights, as obligations of the state to society as a whole, rather than as claims that must be met by the state in each individual case. Programmatic rights, though unenforceable by the individual citizen, can find legal expression, very often in constitutional provisions (Fitzpatrick and Stephens, 2007). One such example, discussed in Chapter 12 in this volume, relates to Tanzania, where the constitution includes a 'right to work', but this 'fundamental' constitutional right is not enforceable through the courts.

Enforceable legal rights

From a legal positivist's point of view, programmatic 'rights' are barely worthy of the name, as captured in the common law maxim 'no right without remedy'. Their interest would lie solely in positive legal rights (sometimes called 'black-letter' rights). In the case of social or welfare rights, such legal rights entitle individual citizens to some defined set of goods or services (a basic income, housing, education), by identifying particular agents (usually, public officials) with 'perfect duties' to fulfil those rights. In cases where legal welfare rights go unfulfilled, individuals can seek redress via the relevant domestic court system or other adjudicative mechanism.

In fact, such legally enforceable entitlements are surprisingly rare with respect to most forms of 'in-kind' welfare, such as education, health and housing, even in the most developed welfare states, though they are somewhat more common with respect to cash transfer benefits (see Dean, 2002). Nonetheless, there are a number of important reasons why enforceable legal rights may be viewed as preferable to purely programmatic rights in addressing issues of social exclusion, marginalisation and disadvantage. First and foremost, they may be considered an important advance on what Goodin describes as 'more odious forms of official discretion' (1986: 232). This perspective is informed both by the welfare rights movement (Dean, 2002) and (relatedly) by theoretical perspectives that acknowledge that welfare interactions between citizens and the state are 'power situations', in which public officials have considerable power over 'claimants' and acknowledge the potentially negative consequences of discretionary or 'charitable' provision. Public officials who administer welfare goods or services have power over claimants because they have an effective sanction against them (Spicker, 1984), and it can be argued that legal rights create a counter-hierarchy of power by giving service users a 'right of action' against service providers (Kenna, 2005). Rights-based approaches can thus be viewed as a form of – albeit state-sanctioned – 'bottom-up' regulation that permits people in situations of little political or economic power a central voice in holding decision-makers to account.

A second and linked argument is that providing goods and services required to meet people's needs as a matter of discretion stigmatises recipients, whereas receiving them as a matter of right does not. When service users are 'beneficiaries' rather than 'rights-holders', there is an implied debt of gratitude as the beneficiary is unable to honour the powerful norm of reciprocity. As such, the giver gains status, and the receiver loses it (Spicker, 1984). There are parallels here with literature that points to the ways in which both the giving and receiving of charity reinforces existing distributions of power and resources, requiring beneficiaries to play the 'role' of the docile, inferior and grateful recipient (Parsell, 2011; Watts, 2013). There are also relevant insights from social psychology that point to the negative consequences of marginalised groups having a 'depressed sense of entitlement' – by preventing members of such groups from perceiving their situation as unjust, existing social inequalities are more likely to be tolerated and perpetuated (Jost *et al.*, 2009).

Rights-based approaches, it is argued, overcome these disempowering and stigmatising impacts of discretionary welfare (Dwyer, 2004), and safeguard the self-respect of welfare recipients (Rawls, 1971/1999) because they reflect their *equal* status as a citizen rather than their unequal status as a dependent (Spicker, 1984). Legal rights thus become a key instrument in supporting a 'politics of recognition' that affords dignity to those living in poverty and using welfare services (Lister, 2004).

While these arguments provide a number of persuasive reasons to support the establishment of legally enforceable welfare rights in tackling social injustice, there are important counter voices.

First, enforceable legal welfare rights may be thought to contribute to the 'juridification of welfare', such that social policy becomes 'over-legalised', frustrating its fundamental purposes (Dean, 2002: 157). According to this critique, legal rights risk imposing bewildering complexity and administrative rigidity on responses to social need and (turning arguments against discretion on their head) stifling the potential for the 'humane and flexible exercise of discretion' (Donnison, 1977: 535; see also Titmuss, 1971; Walker, 2005). In addition, they risk suffusing the delivery of welfare goods and services with a risk-averse and process-orientated culture as a means of avoiding costly legal challenges.

On a broader level, critiques of rights-based approaches have raised concerns that casting individuals as 'unyielding rights-bearers' will encourage them to claim their rights, regardless of genuine need or the impact on others (Kymlicka, 2002). This echoes concerns about the moral hazard associated with welfare rights – that is, that they may encourage dependency and undermine self-reliance (Plant, 2003).

It has also been suggested that seeking to empower marginalised groups with legal rights is misguided on various practical grounds. Goodin (1986), for example, highlights the substantial costs and practical difficulties faced by individuals seeking to challenge the non-fulfilment of their legal rights, arguing that legalistic approaches are fundamentally flawed as they place the burden of responsibility in the wrong place, redistributing power to those people least likely to be able to use it:

> In purely rights-based systems, the rights holders alone have legal standing to complain if officials fail to do their correlative duties. It seems to be sheer folly, however, to make their getting their due contingent upon their demanding it, since we know so well that (for one reason or another) a substantial number of them will in fact not do so.
>
> Ibid.: 255

These arguments focus attention on what is an important – and oft neglected – point about the nature of legal rights. As Bengtsson (2001) emphasises, and contrary to the 'mood' that often surrounds social citizenship discourses, legal rights to welfare goods are inevitably targeted and selective. Indeed, they are increasingly targeted, and not merely at particular categories of people or social risks (e.g. unemployment), but also with respect to a range of 'acceptable' behaviours on the part of recipients (e.g. active job search) (Dwyer, 2004). As such, legal rights considered 'empowering', on the one hand, might also be implicated in coercive processes of social control, on the other (Dean, 2002). Thus, to borrow a legal metaphor, it appears reasonable to conclude that the 'jury is still out' on the relative benefits and disbenefits of 'legalistic' rights as a route to tackling social injustice (see also Fitzpatrick *et al.*, 2014). This is essentially an empirical question, requiring primary research that systematically compares the outcomes and experiences of marginalised groups in national and local contexts with varying degrees of emphasis on legally enforceable rights to welfare goods. That said, such research as does exist on these matters tends to reinforce the case that a framework of enforceable, specific legal rights has beneficial discursive and psycho-social impacts for marginalised groups, with 'statutory rights … becom[ing] internalised as a sense of entitlement' (Lewis and Smithson, 2001: 1477), which in turn helps to construct the claims of these groups as legitimate in the eyes of other stakeholders (e.g. Watts, 2013).

Conclusion

This chapter has attempted to summarise the complexity of the concept of 'rights' at both national and international levels, with a focus on theorising the role that rights-based approaches may play in tackling social injustice, including that experienced by informal economy workers. A key aim has been to emphasise why, if one is endorsing a 'rights-based' approach, it is necessary to be clear about the scope and nature of the approach one is advocating, as natural/human rights, programmatic rights and legal rights all have their distinctive limitations and strengths.

The accounts presented later in this volume make clear that reliance on discourses of universal human rights has not been sufficient to safeguard workers in the informal economy from a range of injustices and disadvantages. Moreover, 'programmatic' constitutional rights to work and to a livelihood (see Chapter 12) have, to date, proven weak in protecting and furthering the interests of informal economy workers. Though some quarters still hold out hope that positive shifts can be secured by asking courts to enforce the abstract rights enshrined in international instruments and some national constitutions, it would seem evident that pursuing progress via nationally negotiated domestic legislation – which establishes specific, concrete and enforceable legal rights – should be at least as important a priority.

Acknowledgements

The chapter draws on Fitzpatrick *et al.* (2014), in which the authors, together with Bo Bengtsson (Uppsala University, Sweden), examine 'rights to housing'. We would like to thank the Editor of *Housing, Theory and Society* for granting permission to draw on this earlier journal article in the current chapter.

References

Alexander, J.M. (2004) Capabilities, human rights and moral pluralism, *The International Journal of Human Rights*, 8(4): 451–69.
Arendt, H. (1963) *On Revolution*, London: Penguin.
Arendt, H. (1973) *The Origins of Totalitarianism*, New York: Harcourt Brace.
Attoh, K.A. (2011) What kind of right is the right to the city?, *Progress in Human Geography*, 35(5): 669–85.
Bengtsson, B. (2001) Housing as a social right: implications for welfare state theory, *Scandinavian Political Studies*, 24(4): 255–75.
Bengtsson, B., Fitzpatrick, S. and Watts, B. (2012) Rights, citizenship, and shelter, in Smith S.J. (ed.), *International Encyclopedia of Housing and Home*, Oxford: Elsevier: 148–57.
Bentham, J. (1789) An introduction to the principles and morals of legislation, in Warnock M. (ed.), *Utilitarianism*, Glasgow: Collins.
Berlin, I. (1969/2002) Two concepts of liberty, in Hardy H. (ed.), *Liberty*, Oxford: Oxford University Press.
Campbell, T.D., Ewing, K.K.D. and Tomkins, A. (2011) *The Legal Protection of Human Rights: Sceptical Essays*, Oxford: Oxford University Press.
Dean, H. (2002) *Welfare Rights and Social Policy*, Harlow: Pearson Education.
Dean, H. (2008) Social policy and human rights: re-thinking the engagement, *Social Policy and Society*, 7(1): 1–12.

Dean, H. (2010) *Understanding Human Need*, Bristol: Policy Press.

Donnison, D. (1977) Against discretion, *New Society*, 41: 534–36.

Dworkin, R. (1977) *Taking Rights Seriously*, London: Duckworth.

Dwyer, P. (2004) Creeping conditionality in the UK: from welfare rights to conditional entitlements?, *The Canadian Journal of Sociology*, 29(2): 265–87.

Finch, J.D. (1979) *Introduction to Legal Theory*, London: Sweet and Maxwell.

Fitzpatrick, S. and Stephens, M. (2007) *An International Review of Homelessness and Social Housing Policy*, London: Department of Communities and Local Government.

Fitzpatrick, S. and Watts, B. (2010) 'The right to housing' for homeless people, in O'Sullivan, E., Busch-Geertsema, V., Quilgars D. and Pleace, N. (eds), *Homelessness Research in Europe*, Brussels: FEANTSA.

Fitzpatrick, S., Bengtsson, B. and Watts, B. (2014) Rights to housing: reviewing the terrain and exploring a way forward, *Housing, Theory and Society*, 31(4): 447–63.

Fortman, B. (2006) Poverty as a failure of entitlement: do rights-based approaches make sense?, *International Poverty Law: An Emerging Discourse*, London: Zed.

Fraser, N. (1989) *Unruly Practices: Power, Discourse and Gender in Contemporary Social Theory*, Cambridge: Polity.

Goodin, R.E. (1986) Welfare, rights and discretion, *Oxford Journal of Legal Studies*, 6(2): 232–61.

Ignatieff, M. (1984) *The Needs of Strangers*, New York: Picador.

Ignatieff, M. (2000) I. Human rights as politics; II. Human rights as idolatry. Paper presented at the Tanner Lectures on Human Values, Princeton University.

Isaac, J.C. (2002) Hannah Arendt on human rights and the limits of exposure, or why Naom Chomsky is wrong about the meaning of Kosovo, *Social Research*, 69(2): 505–37.

Johnsen, S. and Fitzpatrick, S. (2010) Revanchist sanitisation or coercive care? The use of enforcement to combat begging, street drinking and rough sleeping in England, *Urban Studies*, 47(8): 1703–23.

Jost, J., Kay, A. and Thorisdottir, H. (2009) *Social and Psychological Bases of Ideology and System Justification*, New York: Oxford University Press.

Kenna, P. (2005) *Housing Rights and Human Rights*, Brussels: FEANTSA.

Kenna, P. and Uhry, M. (2006) How the right to housing became justiciable in France, https://www.escr-net.org/docs/i/939663, accessed June 2014.

King, P. (2003) Housing as a freedom right, *Housing Studies*, 18(5): 661–72.

Kymlicka, W. (2002) *Contemporary Political Philosophy: An Introduction*, Oxford: Oxford University Press.

Lewis, S. and Smithson, J. (2001) Sense of entitlement to support for the reconciliation of employment and family life, *Human Relations*, 54(11): 1455–81.

Lister, R. (2004) *Poverty*, Cambridge: Polity Press.

Locke, J. (1690) *Two Treatises on Civil Government*, New York: Mentor.

Lukes, S. (2008) *Moral Relativism*, London: Profile.

Mabbett, D. (2005) The development of rights-based social policy in the European Union: the example of disability rights, *Journal of Common Market Studies*, 43(1): 97–120.

Marshall, T.H. (1949) *Citizenship and Social Class: And Other Essays*, Cambridge: Cambridge University Press.

McLachlan, H.V. (1998) Justice, rights and health care: a discussion of the report of the Commission on Social Justice, *International Journal of Sociology and Social Policy*, 18(11/12): 65–84.

Miller, D. (1999) *Principles of Social Justice*, Cambridge, MA: Harvard University Press.

Norman, R.J. (1998) *The Moral Philosophers: An Introduction to Ethics*, Oxford: Oxford University Press.

Nussbaum, M.C. (2000) *Women and Human Development: The Capabilities Approach*, Cambridge: Cambridge University Press.

Nussbaum, M.C. (2011) Capabilities, entitlements, rights: supplementation and critique, *Journal of Human Development and Capabilities*, 12(1): 23–37.

Parsell, C. (2011) Homeless identities: enacted and ascribed, *The British Journal of Sociology*, 62(3): 442–61.

Plant, R. (2003) Citizenship and social security, *Fiscal Studies*, 24(2): 153–66.

Rawls, J. (1971/1999) *A Theory of Justice*, Cambridge, MA: Harvard University Press.

Ross, A. (1974) *On Law and Justice*, London: Stevens.

Scruton, R. (2006) *Political Philosophy: Arguments for Conservatism*, London: Continuum.

Sen, A. (1984) *Resources, Values, and Development*, Cambridge: Harvard University Press.

Sen, A. (1992) *Inequality Reexamined*, New York: Oxford University Press.

Spicker, P. (1984) *Stigma and Social Welfare*, Beckenham: Croom Helm.

Titmuss, R.M. (1971) Welfare 'rights', law and discretion, *The Political Quarterly*, 42(2): 113–32.

Tucker, A. (2015) The anti-democratic turn in the defence of the Human Rights Act. Paper presented at the UK Constitutional Law Association Conference, University of Manchester, 24 June.

Turner, B.S. (1993) Outline of a theory of human rights, *Sociology*, 27(3): 489–512.

UNDP (2000) *Human Development Report*, New York: Oxford University Press.

Waldron, J. (1993) *Liberal Rights: Collected Papers*, New York: Cambridge University Press.

Walker, R. (2005) *Social Security and Welfare: Concepts and Comparisons*, Maidenhead: Open University Press.

Watts, B. (2013) Rights, needs and stigma: a comparison of homelessness policy in Scotland and Ireland, *European Journal of Homelessness*, 7(1): 41–68.

Williams, B. (1995). Ethics, in Grayling, A.C. (ed.), *Philosophy: A Guide through the Subject*, Oxford: Oxford University Press.

4 'Right to the city' and the new urban order

Edésio Fernandes

Summary

The *right to the city* has been a clarion call for a new urban order since the theoretical proposition developed by Henri Lefebvre in the late 1960s. This chapter analyses the role played by the urban-legal paradigm in creating exclusionary patterns of urban development, and argues that a new legal order is an essential underpinning for an inclusive urban agenda based on the right to the city. While many of the debates on the right to the city have emerged from social claims for the right to adequate housing, the principles in this chapter apply equally to legal claims by the urban informal economy.

Introduction

The rapid and intense urbanisation in Latin America over the last 50 years is often contrasted in the literature with an inadequate urban planning system as a way to explain many resulting social problems: high land prices and property speculation, rampant informality, extreme socio-spatial segregation, inadequate urban infrastructure and services, environmental degradation and the like. The literature is largely silent, however, on the role played by national legal systems, which have both contributed to this situation and reacted against it. The pivotal role of the legal order cannot be underestimated (Fernandes and Copello, 2009).

Legal systems have also contributed to informal development in two main ways – through exclusionary legal provisions for registering land and property rights, and through flawed planning systems adopted in many large cities. Both the lack of land regulation and the approval of elitist planning laws that fail to reflect the socio-economic realities that limit access to land and housing by the poor have had a perverse role in aggravating, if not in determining, socio-spatial segregation. Institutional disputes between local and national governments over the power to regulate urban development have also brought about renewed legal problems.

Conflicting legal perspectives have evolved from progressive jurisprudence, the demands of various social movements and a growing legislative debate prompted by divergent stakeholder interests. As a result, legal discussions in Latin America range between anachronistic interpretations of existing legal provisions and a call for a more legitimated and socially responsive legal system. This chapter attempts to expose these tensions and offer some new directions for the debate.

The search for a consistent legal paradigm

In many cities the legal systems regulating urban development are obsolete and inconsistent, resulting in rampant noncompliance and a growing disconnection between the legal and the real city. Important urban management advances promoted by progressive local administrations have often been undermined by obstacles created by outdated national urban-legal orders. Within the broader context of the volatile democratisation processes in Latin America, greater emphasis has been placed on the possibilities that a renewed urban-legal order could advance urban reform. Many academics, politicians, public officials and community organisations understand that the promotion of efficient land markets, socio-spatial inclusion and environmental sustainability will be possible only through the adoption of a clearly defined and consistent new legal paradigm.

Legal principles in general, and particularly those regulating land development rights and property relations, are politically determined and culturally assimilated. Legal systems tend to be complex, as they accommodate different, contradictory and even conflicting provisions adopted over time as a result of evolving socio-political processes. The maintenance of a legal system that does not fundamentally express the realities of the socio-economic and political-institutional processes that it proposes to regulate generates distortions of all sorts.

Making sense of the legal system is a demanding but crucial task that requires the enactment of new laws as well as a consistent effort of (re)interpretation of the principles and provisions in force. However, interpretation may vary significantly according to the legal paradigm adopted by the interpreter. Different paradigms can coexist in the same legal culture, thus bringing about legal ambiguities and potential judicial conflicts, especially in countries where the traditional divide between private law and public law is still unclear.

Three complementary yet competing legal paradigms exist in Latin American countries – *civil law*, *administrative law* and *urban law*. Historically, the hegemonic civilist paradigm, based on a highly partial reading of the *civil codes* and expressing values of classical liberal legalism, has been gradually reformed by the more interventionist paradigm provided by administrative law. A recent, still incipient movement has gone one step further and claimed that only the more progressive framework of urban law would fully provide a comprehensive legal paradigm for contemporary times.

The civil codes and *laissez faire* urban development

The dominant interpretation of the civil codes – as provided by doctrine and jurisprudence and as ingrained in the popular imagination throughout the twentieth century – still tends to overemphasise the rights of owners to the detriment of their responsibilities and fails to consider other social, environmental and cultural interests that result from property ownership. This interpretation gives scarce consideration to use values, since ownership of land and property is conceived largely as a commodity whose economic value is determined mainly by the owner's interests. Longstanding principles of private law, such as the condemnation of all forms of abuse of power and the requirement of a just cause to justify legitimate enrichment, have been largely ignored in this unbalanced definition of property rights.

From this perspective, state action through land management and urban policy is seriously restricted, and major new urban planning initiatives have often led to judicial conflict. Large public projects usually require expensive land expropriation, with the payment of compensation calculated at full market values. Developers' obligations are few and the burden of infrastructure implementation and service provision has fallen largely on the state. While development and building rights are assumed to be intrinsic expressions of individual land ownership rights, there is no established scope for the notion that public administrations should recapture the land value increment generated by public works and services. This legal tradition has been aggravated further by the bureaucratisation of contractual and commercial transactions, as well as by excessive requirements for property registration and access to credit.

Within this individualistic legal tradition, the right to use and dispose of property is often misconstrued as the right not to use or dispose of property. More substantial legal obligations and compulsory orders are virtually non-existent. The prevalence of this paradigm in Brazil, for example, means that while the housing deficit has been estimated as 7.9 million units and people live in some 12 million precarious constructions, another 5.5 million units are empty or underutilised. An estimated 20–25% of serviced land is vacant in some cities.

Also typical of this civil law paradigm is the absolutism of individual freehold to the detriment of collective or restricted forms of property rights such as leasehold or communal, surface and possession rights. Some of these do exist in many civil codes, but they are largely ignored or underestimated. While prescriptive acquisition rights usually require extended periods of land occupation, there is an arsenal of available legal instruments to evict occupiers and tenants.

As a result of this *laissez faire* approach to land development, the urban-legal order in many Latin American cities cannot be considered fully democratic. The process of informal development reflects the reality that more and more people have had to step outside the law to gain access to urban land and housing.

Administrative law and state intervention

Urban planning in some large cities has been supported by the legal principles of administrative law. This public law paradigm has tried to reform the private law tradition, but it has limited the scope of the notion of the 'social function of property'. Since the 1930s, this concept has existed in most national constitutions as a nominal principle. This more interventionist paradigm recognises the state's 'police power' to impose external restrictions and limitations on individual property rights in the name of the public interest, thus supporting traditional forms of regulatory planning.

These have been timid attempts, however, because the imposition of legal obligations, compulsory orders and requirements of land reservation still tend to be met with strong popular and judicial resistance. In most countries, the courts have ruled that the state can impose certain limitations on property rights, but the imposition of obligations on landowners and developers has been more difficult. This is particularly the case with local laws that have tried to determine the earmarking of land or units for social housing as a condition for the approval of the development and have been declared unconstitutional.

Many cities continue to approve new land subdivisions, even though they already have a large stock of vacant plots. The problem is that they do not have the legal instruments to determine their use according to a social function. While developers have been held more responsible for the implementation of infrastructure, some enormous developments, including high-income gated communities, have been approved without reserving land or housing units for domestic and service workers. This results in new informal developments and the greater densification of older settlements to accommodate the low-income sector.

In some cities that have attempted zoning, master plans and other complex urban laws, a tradition of bureaucratic planning has emerged that reflects little understanding of how urban and environmental regulation impacts the formation, and increase, of land prices. Urban planners still have difficulties challenging the established notion that land and property owners have automatic rights to the gains in value resulting from urban planning and development. Most public administrations have not recaptured the generous land value increment generated by public works and services, or by changes in urban legislation governing use and development rights.

Most planning systems have failed to recognise the state's limited capacity to act so as to guarantee the enforcement of urban legislation. As a result, plans have not been properly implemented, and many forms of disrespect for the legal order have been left unquestioned. In some cities, it takes years to license important development processes such as land subdivisions, which also affects the process of informal development.

Another recurrent problem is the parallel, sometimes antagonistic development of distinct urban and environmental legal orders, with environmental provisions frequently being used to oppose socially oriented housing policies. In socio-political terms, most planning laws do not involve substantial popular participation in either their formulation or implementation.

By failing to change the dynamics of land markets, supposedly 'contemporary' planning policies often end up reinforcing traditional processes of land and property speculation and socio-spatial segregation. Urban planning has often been inefficient in promoting balanced land development and, instead, has benefited land developers, property investors and speculators. Their profits have been maximised by the significant growth in prices resulting from urban regulations that determine urban development boundaries. The areas left for the urban poor are those not regulated for the market, such as public land and environmentally sensitive areas.

It is from this tension between the interpretation of civil codes and bureaucratic planning laws that informal development and socio-spatial segregation have resulted – law has been one of the main factors determining urban illegality. In cases where significant attempts have been made to promote socio-spatial inclusion and environmental sustainability, the urban-legal order still fails to fully support the prevailing practice of urban management.

For example, aspects of public–private partnerships and the involvement of NGOs in the provision of public services have been questioned because of confusion between private and public values. Nominally recognised social rights, such as the right to adequate housing, have not been fully enforced due to the lack of necessary processes, mechanisms and instruments.

The 'right to the city'

An agenda which has underpinned a fundamental challenge to the dominance of traditional processes of land development and property rights has evolved from the work of French Marxist philosopher and sociologist Henri Lefebvre, and his arguments for a new form of citizens' rights, termed the 'right to the city', based on a conceptualisation of the socio-spatial function of property.

Although Lefebvre has received some recognition of late, especially through the works of English-speaking authors, the full extent of his fundamental contribution to urban research still needs to be properly appreciated. A growing number of studies in the English language have explored the notion of the right to the city from a socio-political perspective (Harvey, 1973; Mitchell, 2003).

In France, Lefebvre's work was partly eclipsed by the enormous political and academic influence of Manuel Castells's *La question urbaine* (1972). Some of Lefebvre's books (for example, *La révolution urbaine* (1970) and, especially, *La production de l'espace* (1974)) are still fundamental sources of sharp insights and enlightening analysis on crucial socio-economic, political and cultural aspects of the process of urban development, particularly the dynamics of the urban land market. Some of his ideas were even visionary: long before the concept became widespread, Lefebvre was talking about the phenomenon of globalisation (*mondialisation*).

Although the concept of the right to the city was first proposed in two of his books, *Le droit à la ville* (1968) and *Espace et politique* (1973), Lefebvre's most consistent elaboration on the nature of this concept can be found in a lesser-known book, *Du contrat de citoyenneté* (1990), in which the original concept was further developed and given more socio-political content, and in which the importance of reforming the longstanding liberal tradition of citizenship rights was strongly argued.

Indeed, more than 200 years have gone by since the Declaration of the Rights of Man and Citizens on 26 August 1789 passed by France's National Constituent

Assembly – the touchstone of the democratic order – was approved. Throughout this period, as a result of wars, political struggles and social movements the set of human rights originally affirmed has gradually been recognised, expanded and incorporated into the ordinary lives of individuals and social groups.

However, despite the profound socio-economic, political and territorial changes that have taken place throughout the past two centuries, particularly those promoted by the processes of industrialisation and urbanisation, the same cannot be said about the set of rights of citizens, which remain, to a large extent, the same political rights originally stipulated in 1789. Thus, Lefebvre argued that updating the Declaration of the Rights of Citizens is of utmost importance so that new legal-political conditions can be created so as to affirm a notion of social citizenship.

In his work, Lefebvre suggested some of these interrelated political rights still need to be fully recognised: the right to information; the right of expression; the right to culture; the right to identity in difference and in equality; the right to self-management, that is, the democratic control of the economy and politics; the right to public and non-public services; and, above all, the right to the city.

The right to the city would consist of the right of all city dwellers to fully enjoy urban life with all of its services and advantages – the right to habitation – as well as taking direct part in the management of cities – the right to participation. In other words, Lefebvre stressed the need for the full recognition of *use values* in order to redress the historical imbalance resulting from the excessive emphasis on *exchange values* typical of the capitalist production of the urban space. This vital link between cities and citizenship has become an imperative given the escalating urbanisation of contemporary society at a global level.

However, revealing and exciting as his ideas are, the fact is that Lefebvre's concept of the right to the city itself was more of a political-philosophical platform and did not directly explore how, or the extent to which, the legal order has determined the exclusionary pattern of urban development. To Lefebvre's socio-political arguments, another line of arguments needs to be added – that is, *legal* arguments leading to a critique of the legal order not only from external socio-political or humanitarian values, but also from within the legal order.

Such a full understanding of the crucial role played by the legal order is the very condition for the promotion of a profound legal reform, which in turn is the condition for the promotion of urban reform leading to social inclusiveness and sustainable development.

It is in this context that the Latin American experience, and particularly the case of Brazil, needs to be better understood. Perhaps more than anywhere else in the world, Lefebvre's concept of the right to the city has been extremely influential in Latin America; since the mid-1970s a consistent socio-political mobilisation has tried to realise it in both political and legal terms.

Urban law and the principles of legal reform

There are several guiding principles which are common to the socio-political processes claimed for urban law reform, and which have been incorporated in several national constitutions and legal orders of Latin America. First is that of the socio-environmental function of property and of the city – an expression of the broader principle according to which the regulation of urban development is a public matter that cannot be reduced to either individual or state interests – and the intertwined collective rights that include: the right to urban planning; the social right to housing; the right to environmental preservation; the right to capture surplus value; and the right to the regularisation of informal settlements.

Second is the indivisibility of urban law and urban management. This principle has been expressed through three integrated processes of legal-political reforms, namely the renewal of representative democracy through the recognition of the collective right to participation in urban management; the decentralisation of the decision-making processes, especially local government; and the creation of a new legal administrative framework to provide more clarity to the new relations taking place between state and society.

In other words, this growing socio-political movement of legal reform in Latin America has been based on the two pillars Lefebvre proposed as the core of the right to the city, namely the *right to habitation* and the *right to participation.*

The proponents of urban law have argued that it is possible, and indeed necessary, to look in the civil codes for principles that allow for strong legal arguments to support sound state intervention in, and social control of, the regulation of land and property-related processes. The reinterpretation of traditional legal principles, as well as the emphasis on neglected principles (such as the notion of no legitimate enrichment without a just cause), can help to enable significant progress in the formulation of urban land policy.

This effort requires sophisticated legal expertise, as it potentially involves legal debates and judicial disputes whose results are far from certain. From the viewpoint of the urban communities and public administrations committed to promoting inclusive policies, this approach seeks to organise the overall regulatory framework, in part through the enactment of new laws that more clearly express the principles of urban law.

The right to the city in Brazil

In Brazil, profound changes took place following the 1985 transition to civilian rule. The promulgation of the 1988 Federal Constitution paved the way for the progress of the legal reform movement, especially by recognising collective rights, by affirming the central role of local government and also by declaring that representative democracy is to be reconciled with participatory political process (Fernandes, 1995). Throughout the 1990s, several Brazilian municipalities started to effect the constitutional provisions and principles in their own redefined legal-urban orders, and Brazil became a laboratory of sorts for new strategies of local governance and direct democracy; in particular, the ground-breaking experience of the participatory budgeting process was introduced in some cities.

However, there were still legal controversies over the new constitutional provisions, and conservative legal arguments were formulated in order to undermine the innovative local political-institutional strategies of urban management. For this reason, throughout the 1990s the social movements and NGOs, under the umbrella of the National Forum of Urban Reform, kept pressing for the National Congress to approve of a federal law governing urban development and policy, regulating the constitutional chapter and thus clarifying the outstanding legal problems.

Thus, on 10 July 2001, a ground-breaking legal development took place in Brazil with the enactment of Federal Law no. 10.257, entitled the City Statute (*Estatuto da Cidade, Lei no 10.257*) which aims to regulate the original chapter on urban policy introduced by the 1988 Federal Constitution and in which the right to the city is explicitly recognised as a collective right.

The City Statute confirmed and widened the fundamental legal-political role of municipalities in the formulation of directives for urban planning, as well as in conducting the process of urban development and management. The City Statute deserves to be known at the international level because it is an inspiring example of how national governments can effect the principles and proposals of the UN-Habitat global campaigns on Urban Governance and on Secure Tenure for the Urban Poor. It is impossible to underestimate the impact the new law is having on Brazil's legal and urban order.

The City Statute has four main dimensions, namely conceptual, providing elements for the interpretation of the constitutional principle of the social function of urban property; providing new legal, urban and financial instruments to municipalities for the construction and financing of a different urban order; indicating new processes for the democratic management of cities; and identifying legal instruments for the comprehensive regularisation of informal settlements in private and public urban areas. Combined, these dimensions provide the content of the right to the city in Brazil, as well as indicating the conditions for the materialisation of the new social contract proposed by Lefebvre.

In conceptual terms, the City Statute broke with the longstanding, individualistic tradition of civil law and set the basis of a new legal-political paradigm for urban land use and development control in Brazil, especially by consolidating the constitutional approach to urban property rights – namely the right to urban property is assured and recognised as a fundamental individual right provided that a socio-environmental function is accomplished. Moreover, it is particularly the task of municipal governments to control the process of urban development through the formulation of territorial and land use policies in which the individual interests of landowners necessarily coexist with other social, cultural and environmental interests of other groups and the city as a whole. For this purpose, municipal government was given the power to determine the balance between individual and collective interests over the utilisation of this non-renewable resource essential to sustainable development in cities – that is, urban land. All Brazilian municipalities with more than 20,000 inhabitants had to approve their master plans by the end of 2006.

In Brazil, the combination of traditional urban planning mechanisms – zoning, subdivision, building rules and so on – with the new instruments introduced – compulsory subdivision/edification/utilisation order; extra-fiscal use of local property tax; expropriation sanction with payment in titles of public debt; surface rights; preference rights for the municipality; onerous transfer of building rights and so on – has opened a new range of possibilities for the construction and financing of a new urban order which is, at once, economically more efficient, politically fairer and more sensitive to social and environmental questions.

Another fundamental dimension of the City Statute concerns the need for municipalities to integrate urban planning, legislation and management so as to democratise the local decision-making process. Several mechanisms were recognised to ensure the effective participation of citizens and associations in urban planning and management: executive power (consultations, creation of councils, committees, referendums, reports of environmental and neighbourhood impact and, above all, the practices of the participatory budgeting process), legislative power (public audiences, popular initiative to propose bills of urban laws) and the judiciary (civil public action to protect the legal-urban order, and a collective law). This approach emphasised the importance of establishing new relations between the state, private and the community sectors, especially through partnerships and urban/linkage operations to be promoted within a legal-political and fiscal framework, the principles of which are still to be more clearly defined.

Last, but not least, the City Statute also recognised the collective right to the regularisation of informal settlements on public or private land, and improved the legal instruments to enable municipalities to promote land tenure regularisation programmes and thus democratise the conditions of access to urban land and housing. As well as regulating the constitutional institutes of special *usucapião* rights (adverse possession rights) and 'concession of the real right to use' (a form of leasehold) to be used in the regularisation of informal settlements in both private and public land, the new law went one step further and admitted the collective utilisation of such instruments.

Taking forward the new urban order

Although this process is more advanced in Brazil (mainly through the 1988 Federal Constitution and 2001 City Statute) and Colombia (mainly through the 1991 Constitution and Law no. 388/1997), a series of common principles have been incorporated into the legal orders of other Latin American countries (Fernandes, 2007a, 2007b; Copello and Mercedes, 2003, 2007).

This chapter argues that enshrining the right to the city in law, rather than as a social contract through a citizen charter, is critical to establishing the new urban order. The first most important structural principle is the notion of the social function of property, including public property and property registration. Cities result from a collective process, and the promotion of a balanced territorial order is at once a collective right and the obligation of the state. The urban order cannot be determined exclusively by the individual rights and interests of landowners, nor only by state interests. Public intervention should be promoted through administrative limitations on property rights, and through legal responsibilities and development requirements.

Related legal principles operate to: determine that a just distribution of the costs and opportunities of urban development is promoted between owners, developers, the state and society; affirm the state's central role in determining an adequate territorial order through the planning and management system; establish a clear separation between property and development/building rights; determine different criteria for the calculation of compensation in different expropriation and other contexts; reduce the required time for adverse possession to take place for the materialisation of social housing, and recognise more strongly the rights of occupiers and tenants.

A whole range of collective rights guides the processes of land use and development, such as the rights to urban planning, adequate housing and a balanced environment; the community's right and state's obligation to recapture the land value increment generated by state action and urban legislation; and the right to the regularisation of consolidated informal settlements.

Some Colombian cities have amassed significant financial resources through land value capture mechanisms, making it possible (if not always feasible) to formulate a more sustainable process of legal access to serviced land by the urban poor. In Brazil, some municipalities also have been able to generate impressive financial resources as a result of 'urban operations', in which development and building rights are negotiated within the framework of a master plan. Regularisation programmes involving both the upgrading and legalisation of consolidated settlements also have been promoted in several countries.

However, the dispute among legal paradigms continues, and all new principles and rights are still the subject of fierce debate. Colombia's Constitutional Court has consistently adopted a progressive interpretation, sustaining the notions of the social function of property and the social right to housing. A recent study of judicial decisions by high courts from several Brazilian states showed that the new legal paradigm has been assimilated in some 50% of those decisions, with the conservative civil code paradigm still orienting the other decisions (Mattos, 2006).

In many countries progressive jurisprudence has been restricted by the strong tradition of positivism and formal legalism, which still views the law merely as a technical tool to resolve conflicts, as if it were totally independent from socio-political and economic processes. Most judges observe the civilist paradigm, which is taught in anachronistic law school syllabuses. Progressive decisions by local judges often are revoked by more traditional higher courts.

The second important structural principle of this emerging urban-legal order is the integration of law and management within the framework of three intertwined legal-political changes:

- restoration of local democracy, especially in Brazil, through the recognition of several forms of popular participation in law-making (as a condition for the legitimacy of new urban laws and their legal validity) and in urban management (as in the participatory budgeting process);

- decentralisation of decision-making processes by strengthening local administrations, addressing the need for a metropolitan level of policy-making and action, and articulating intergovernmental systems to overcome accumulated urban, social and environmental problems, and
- creation of a new set of legal references to give more support to the new relations being established between state and society, particularly through public–private partnerships and other forms of relationships of between the state and the private, community and voluntary sectors.

Whatever the shortcomings of the process, the enormous challenge put to countries and cities promoting urban law reform is to guarantee the full enforcement of the newly approved laws.

The construction of a new urban-legal order in Latin America and other regions is an evolving debate full of contradictions and challenges, and none of the recent developments can be taken for granted. If the greater politicisation of urban law-making has created a widened scope for popular participation in the process to defend collective rights and social interests, for the same reason the new laws have generated increasing resistance on the part of conservative stakeholders.

The full implementation of the possibilities introduced by the new urban-legal order in Brazil, Colombia and elsewhere will depend on several factors, but above all on the renovation of the processes of socio-political mobilisation, institutional change and legal reform.

Acknowledgements

This chapter was compiled from two published papers: Fernandes, E. (2007b) and Fernandes, E. and Copello, M. (2009). The author thanks the editors of *Social and Legal Studies* and *Land Lines* for permission to republish, and María Mercedes Maldonado Copello for her advice and inputs on one of the original papers from which this work was drawn.

References

Castells, M. (1972) *La question urbaine*, Paris: François Maspéro.
Copello, M. and Mercedes, M. (2003) Reforma territorial y desarollo urbano – Experiencias y perspectivas de aplicacion de las leys 9 de 1989 y 388 de 1997, Bogota: CIDER-Universidad de los Andes/LILP/FEDEVIVIENDA.
Copello, M. and Mercedes, M. (2007) El proceso de construcción del sistema urbanístico colombiano: entre reforma urbana y desarrollo territorial, in Fernandes, E. and Alfonsin, B., *Direito Urbanístico: Estudos brasileiros e internacionais*, Belo Horizonte, Brazil: Del Rey.
Fernandes, E. (1995) *Law and Urban Change in Brazil*, Aldershot: Avebury.
Fernandes, E. (2007a) Implementing the urban reform agenda in Brazil, *Environment and Urbanization*, 19(1): 177–89.

Fernandes, E. (2007b) Constructing the 'right to the city' in Brazil, *Social and Legal Studies*, 16(2): 201–19.

Fernandes, E. and Copello, M. (2009) Law and and policy: shifting paradigms and possibilities for action, Lincoln Institute of Land Policy, *Land Lines*, July.

Mattos, L.P. (2006) *Função social da propriedade na prática dos tribunais*, Rio de Janeiro: Lumen Juris.

5   Reclaiming space

Street trading and revanchism in Latin America

*Peter Mackie, Kate Swanson
and Ryan Goode*

Summary

Public space is core to the image of the city in a globalising world, and street trade is seen as anathema to well-planned, functioning and attractive cities, hence policies of spatial displacement are common. Drawing on Neil Smith's ideas of 'revanchist urbanism', the chapter discusses the impacts of displacement on street traders but also, importantly, examines street trader resistance and the emergence of more tolerant, supportive policy environments. The focus of this chapter is Latin America, as it is here that the more tolerant post-revanchist turn has been documented.

Introduction

This chapter examines the battle over public space between street traders and those who seek their removal. It argues that long-established zero-tolerance policies of street trader clearances persist today; so-called *revanchist urbanism* whereby urban space is claimed for elites to exclude the poor. We explore the known impacts of these policies before examining emerging, more tolerant *post-revanchist* approaches, which recognise traders' needs to appropriate public space to maintain their livelihoods.

Much street trading takes place within public space, which is subject to laws, directives and bylaws relating to highways, public order or urban management. Many of these are conflicting and make it impossible for street traders to trade legally, as discussed in Chapter 1 and elsewhere in the book. In operational terms, public space is managed by the police, municipal authority officials and other enforcers, who may function very differently under varying political regimes or leadership, even in similar legal contexts. This chapter explores how different policy approaches can result in very different outcomes for traders. The concept of public space is here taken to include 'formal public space in parks, squares and streets, and space at the margins ... where public access is possible but not formalised' (Brown, 2006: 22), but is broadened to include space within buildings appropriated by traders and opened to public access.

Street trading has long been seen as anathema to well-planned, functioning and attractive cities. Consequently, there has been a proliferation of policies which have sought the removal of street traders from city centres across the globe. The nature of removal policies has varied in relation to the force used, and the extent and location of any alternative trading sites provided, and yet the agenda remains the same: street traders should be removed from public spaces in favour of more dominant social groups. These processes of trader displacement have been well documented across the globe; for example, in Africa (Musoni, 2010), Latin America (Bromley and Mackie, 2009; Crossa, 2009; Hunt, 2009; Peña, 1999; Swanson, 2007) and Asia (Turner and Schoenberger, 2012). Drawing on Neil Smith's seminal work in New York (1996), several academics have described this policy paradigm affecting street trading as a form of *revanchist urbanism* (Mackie *et al.*, 2014; Swanson, 2007), whereby dominant social groups are reclaiming the city's streets from marginalised street traders. More recently, in the Latin American context, there has reportedly been a shift towards a '*post-revanchist*' turn (Galvis, 2014; Mackie *et al.*, 2014), whereby once revanchist urban policies have become more supportive and tolerant of traders.

Despite the proliferation of individual studies on street trader displacement, and the two Latin American studies which report a post-revanchist turn, to date there has been no attempt to synthesise the key findings in order to draw generalisations about the impacts of revanchist or post-revanchist policies on street traders. By examining the known impacts of these policies, key lessons can be identified for future urban policy-making and street trade. This chapter will focus primarily on findings from Latin American research, where post-revanchism and studies of displacement have been widely documented. However, there are abundant similarities in trader displacements in other regions, and lessons are relevant elsewhere.

The evidence in this chapter is based on a review of key literature. The chapter uses the 'battlegrounds framework', developed by Mackie *et al.* (2014), to analyse documented conflicts between street traders and those who manage public spaces. The battlegrounds framework consists of four domains, all of which are explored: spatial, political, economic and socio-cultural.

The chapter is divided into three sections. The first section introduces the *concepts of revanchism* and *post-revanchism*; the second section uses the battle-grounds framework to examine the *impacts of revanchism on street trading* and the way in which political, economic and cultural claims underpin street traders' access to public space; and the final section explores emerging evidence on the *impacts of a post-revanchist turn*. The conclusions identify key lessons for the future of urban policies targeted at street trading.

Revanchist urbanism

Revanchism is a concept that emerged in response to repressive urban policies in New York City. In the 1990s, New York was perceived as a city out of control, as gangs, drugs and crime ruled the streets. The Mayor, Rudolph Giuliani, pledged to tackle New York's high rates of crime and disorder by focusing on petty crimes and so-called street nuisances. Working closely with Police Commissioner William Bratton, New York launched what came to be known as 'zero-tolerance' policing. This punitive model of policing has since been adopted by cities around the world, particularly in Latin America.

Giuliani and Bratton drew inspiration for zero-tolerance policing from a theory put forth by criminologists James Wilson and George Kelling. Dubbed the 'broken windows theory', the authors speculated that lesser offences such as graffiti, abandoned cars, broken windows and other 'untended behaviour' in cities would lead to an overall 'breakdown of community controls' (Wilson and Kelling, 1982: 3). Specifically, they argued that:

> Social psychologists and police officers tend to agree that if a window in a building is broken and is left unrepaired, all the rest of the windows will soon be broken. ... One unrepaired broken window is a signal that no one cares, and so breaking more windows costs nothing.
>
> Ibid.: 2–3

Because broken windows give the impression that 'no one cares', the authors argued that officers must target these sorts of lesser offences in order to deter more serious crimes. Drawing from these assumptions, this largely speculative theory became the inspiration for zero-tolerance policing strategies.

Giuliani and Bratton believed that the key to combating crime was to target anti-social behaviour or 'quality of life' crimes. Together they crafted a document entitled, *Police Strategy No 5: Reclaiming the Public Spaces of New York*. The document outlined a list of unruly behaviours and targets of 'broken windows policing', which included: squeegee cleaners, boomboxes, graffiti, panhandling, reckless bicyclists, street artists, street vendors and 'dangerous mentally ill street people' (Smith, 1998: 5). According to Giuliani, 'disorder in the public spaces of the city' presents 'visible signs of a city out of control' (ibid.: 3). And so began the crackdown to rid New York City's public spaces of urban disorder.

Yet, these policies were met with much criticism, particularly due to an increase in human rights abuses at the hands of the New York City police. One critic was urban geographer Neil Smith (Smith, 1998) who, describing the increasingly punitive policies of New York, drew parallels to late nineteenth-century France, when the bourgeoisie sought to restore order on city streets and take revenge on those who had tarnished their morally conservative vision. Smith argued that New York City's urban policies also represented a punitive, vengeful, right-wing reaction by the upper-middle classes over the supposed theft of the city. Drawing from the word *revanche* – French for revenge – zero tolerance was described as a 'revanchist' attempt to reclaim the city for the elite. Smith further argued that it represented:

> a vendetta against the most oppressed – workers and 'welfare mothers', immigrants and gays, people of color and homeless people, squatters, anyone who demonstrates in public. They are excoriated for having stolen New York from a white middle class that sees the city as its birthright.
>
> Ibid.: 1

Smith illuminated an emerging class struggle over the basic right to the city. The struggle was not only about access to public space, but also about real estate, investment, property and profits. As the streets were swept clean of undesirables, property investors moved in to gentrify working-class neighbourhoods and displace long-term residents. The wider the rent-gap, the higher the potential profit for investors. Smith likened the situation in New York City to the conquest of the American Wild West. Much like the white settlers who displaced Native Americans for access to western territories, white urban elites displaced the working-class and black and minority ethnic (BME) populations in New York's Lower East Side.

Within a short period of time, the administration heralded zero-tolerance policing as a success. Crime had fallen and the city's street youth, traders and homeless had been swept away. Investment began to flow in and Giuliani and Bratton were lauded as saviours of a city renowned for being one of the most violent in America. Yet, the effectiveness of zero tolerance on reducing crime rates has been widely debated (i.e. Bowling, 1999; Harcourt and Ludwig, 2006). Evidence suggests that the 1990s crime reductions in New York City may have been more influenced by nationwide reductions in crack use, and a demographic shift, than zero-tolerance policing strategies. Nevertheless, global politicians and the media took note and Bratton and Giuliani began to export their zero-tolerance policing model around the world. Latin America was of particular interest given its high urban crime rates and was seen as 'the new frontier of reform for police work' (Bratton and Andrews, 2001).

Some scholars have questioned the usefulness of revanchism as a concept. DeVerteuil (2014), for instance, argues that this punitive framework is often too one-sided and fails to grasp the complexities of urban policy implementation. He argues that the revanchist framework tends to focus on 'singularly punitive, sometimes exaggerated and polemical descriptions' (ibid: 875) at the expense of more balanced views. Others suggest that in some regions of Latin America there may be a post-revanchist turn (Galvis, 2014; Mackie *et al.*, 2014), where cities have adopted more sympathetic approaches to the visible signs of poverty on their streets.

The next section discusses specific examples from Latin America, some that illustrate punitive revanchist policies in Brazil, Colombia, Ecuador, Guatemala, Mexico and Peru and others that show more tolerant, post-revanchist approaches in Colombia and Peru.

Revanchist urbanism and street trading in Latin America

Drawing on a review of key literature on the Latin American region we examine how revanchist urbanism has affected street traders over recent decades. In some cities, the impacts of revanchism and zero-tolerance policing on street traders have been extremely damaging, with profound spatial, economic, political and socio-cultural implications. In what follows we use the battlegrounds framework

(Mackie *et al.*, 2014) to explore the impacts relating to each of these areas in revanchist Latin America cities.

Spatial displacement

The high degree of spatial inequality in Latin America is imprinted on the landscapes of its major cities. Street traders often work in contested public spaces that are near houses of the very wealthy, international tourist zones, or areas targeted for real-estate/capital investments. Zero-tolerance policies tend to focus on these contested spaces, forcing either negotiation, confrontation, or outright displacement of street traders.

Mexico City has witnessed a drastic transformation of its historic centre (Crossa, 2009; Davis, 2013; Mountz and Curran, 2009, Becker and Müller, 2013). In 2002, Giuliani was contracted by billionaire Carlos Slim Helú (at a price of US$4.3 million) to guide the city through its own version of zero-tolerance policing. While drug traffickers and arms traders centred their activities in the urban core, the new policy focused on so-called 'quality of life' crimes, such as street trading, squeegee windscreen-cleaning and car guarding. Under zero tolerance, the number of punishable minor infractions increased significantly, along with fines and prison sentences. Within the first week of the new Civic Culture Law (*Ley de Cultura Cívica del Distrito Federal, 2002*), up to 340 informal parking guards (*franeleros*) were arrested per day (Becker and Müller, 2013: 83). In doing so, the working poor were forcibly displaced from the streets.

A similar law, the *Código de Posturas*, was introduced in 2010 in Belo Hoizonte, setting out rules relating to the use of public space and including an end to street trading. However, unlike the Mexico City example, this law was accompanied by the construction of an alternative trading space in a new shopping district. According to Carrieri and Murta (2011), thousands of street traders were moved from the streets and into buildings in the new shopping district. Similar examples of street trader relocations from central public spaces and key thoroughfares to market buildings have been documented across Latin America (Bromley and Mackie, 2009; Donovan, 2008; Meneses-Reyes and Caballero-Juarez, 2014).

In response to enforced displacement, and motivated largely by the need to maintain livelihoods, street traders have often remained in or returned to busy public spaces. In Mexico City, street traders devised creative strategies to evade the police. Crossa (2009) describes how traders engage in *torear*, which figuratively means to 'deceive' or 'tease'. These traders work in prohibited zones of the city by remaining mobile and employing 'lookouts' to notify them of approaching police. Typically, *toreros* remain mobile by carrying their goods, keeping them strapped to their bodies, or placing them on blankets that can be quickly removed. Mexico City authorities have responded to the success of *toreros* by enforcing harsher penalties, including jail time, for any trader caught vending in prohibited zones. In Cusco, Bromley and Mackie (2009) document how street traders adopted an alternative form of spatial resistance to planned relocation policies. Street traders chose to rent space in courtyards of old colonial buildings and established small informal markets, abandoning or resisting relocation to decentralised market buildings.

The consequences of forced displacement, including those where alternative trading sites are offered, are explored in relation to the political, economic and socio-cultural battlegrounds.

Political displacement

Political struggles also take place, as street traders deploy various strategies to secure power and access to urban space, and formal trading associations have become common across Latin America to help traders secure legal rights to the city. Trading associations have achieved some successes, most often in securing alternative trading locations from local government after a relocation decision has been made (Bromley and Mackie, 2009); however, in these new locations we often see the demise of associations and collectives. In the absence or failure of effective formal unions, street traders have relied on other tactics to have their voices heard and gain legitimacy in urban public spaces, ranging from protests to petitions (Crossa, 2009, 2013).

The decision to relocate street traders rarely emerges from the concerns and wishes of the traders. In Antigua Guatemala, between 2002 and 2003 handicraft vendors were relocated from the streets to a new market, with limited consideration given to the needs or the views of the traders (Little, 2005) and, in Colombia, Hunt (2009) made similar observations, where policies to remove traders from public spaces were decided upon and only then were tokenistic negotiations entered into. It appears that once a decision to remove traders from the streets has been made, the political organisation of street traders becomes more influential on the nature of the displacement. In Guatemala, Little (2005: 87) describes how vendors reluctantly left the streets and were not 'satisfied with the move or the location', but he recognises that their strength as organised political actors at least ensured an alternative trading space was offered. Bromley and Mackie (2009) reached a similar conclusion in Cusco, where street traders were only relocated to decentralised markets if they had licences and those affiliated to particular associations received more favourable locations.

Physical displacement can result in political displacement (Mackie *et al.*, 2014). In Belo Horizonte, the relocation of street traders to buildings in the new shopping district led to 'the fragmentation of the informal alliance' that existed between traders (Carrieri and Murta, 2011: 222). While on the streets, traders supported each other by watching over each other's stalls, supplying goods at short notice and warning each other of potential inspections, but Carrieri and Murta (2011) found that this alliance withered in the new building.

While municipal authorities are clearly in a position of power relative to street traders and displacements and relocations are commonplace, there are many instances of concerted trader resistance. Crossa (2013) documents a fascinating example of resistance in Mexico City, which she describes as a 'playful protest' against a Recovery of Public Spaces policy (*Recuperacion de Espacios Publicos*). The policy aimed at refurbishing historic buildings, upgrading infrastructure and improving safety and security by heightening police presence. Crossa (ibid.) focuses on one particular plaza where this policy resulted in the removal of hundreds of street traders. At the onset of the works, street traders and different organisations came together to form a peaceful sit-in (*planton*), displaying photographs, banners and videos of the struggles streets traders and other artisans had faced. The power struggle between street traders and authorities was ongoing when Crossa (ibid.) was writing, but, irrespective of the outcome, it provides an excellent example of the sort of action traders are taking to attempt to subvert power relations with authorities.

Another example of street trader resistance is petitioning. Little (2005) explains that this has been the most common form of resistance in Guatemala, where traders have petitioned for relatively modest demands such as permits to sell, the creation of centrally located markets and improved security. Interestingly, Little (ibid.) claims that traders seek signatures from tourists as they believe this invokes sympathy, while also giving their demands greater credence with the mayor. It is also notable that Little (ibid.) suggests street traders do not expect changes to be made by local authorities; instead, a petition is measured as a success if it results in reduced harassment by the authorities.

Economic displacement

The economic impacts of revanchist urban policies are significant and long-lasting for street traders. As traders are displaced from the streets, they lose their main source of income, which can have devastating impacts on individuals and families. This situation has been well documented in Cusco (Bromley and Mackie, 2009), Quito and Guayaquil (Swanson, 2010), Guatemala City (Véliz and O'Neill, 2011), Mexico City (Crossa, 2009) and elsewhere.

In Cusco, where many street traders were relocated to decentralised market buildings, the main economic impact was on the reduction in sales, which forced traders into poverty (Mackie *et al.*, 2014). Sales reduced because many of the new markets were barely functioning and those who returned to the streets, in an effort to maintain their livelihoods, faced a heightened police presence, which made sales more challenging. Problems in new market buildings were also reported by Carrieri and Murta (2011) in Belo Horizonte; they explained that strict opening hours leave little flexibility if insufficient sales are made – irrespective of whether the trader makes an income they must pay rent. While some of the problems faced in Cusco are clearly shared in Belo Horizonte, Carrieri and Murta (ibid.) report that traders also reported benefits in the new buildings, largely focused on the physical conditions and protection from the weather.

Where street traders are not offered an alternative trading location, or the alternative location fails to provide a livelihood, street traders are forced into

Figure 5.1 Urban elites displace the vulnerable: market traders in Quito, Ecuador.

alternative coping strategies. In Ecuador, street traders have turned to transnational migration as an alternative means of income. The streets of Quito and Guayaquil were swept clean through urban redevelopment schemes beginning in the early 2000s. This directly affected the livelihoods of indigenous street traders, particularly the Kisapincha from the central Andes, many of whom survived solely on their street earnings (Swanson, 2010). Given few other employment options within Ecuador, a number of Kisapincha chose to migrate to New York City, at great cost, both economically and personally. Young women and men leave their children behind and risk their lives migrating through Central America and Mexico. For those who make it to New York City, the streets remain a familiar employment option, but instead of working on the streets of Ecuador, they now stand in front of large supermarkets in Brooklyn.

Socio-cultural displacement

Human and cultural rights of street traders are often abused through policies of removal and displacement. The frequent use of violence to remove traders is perhaps the most obvious human rights violation but traders also express concerns about the impacts of displacement on their longstanding and important social networks.

Mexico City has witnessed formal Equality and Human Rights Commission (EHRC) complaints due to its zero-tolerance policing, particularly against police abuse directed at sex workers, car guarders and squeegee windscreen cleaners (Beckett and Müller, 2013). According to Wacquant (2003), police violence in Brazil stems from historically constructed race and class hierarchies that derive from slavery and an agrarian past which is:

> backed up by a hierarchical, paternalistic conception of citizenship based
> on the cultural opposition between *feras e doutores*, the 'savages' and the

'cultivated', which tends to assimilate marginals, workers and criminals, so that the enforcement of the class order and the enforcement of public order are merged.

Ibid.: 199

Others examine the impacts of clearance policies on families and communities. Guatemala City has undergone an urban redevelopment scheme directed at sweeping street traders from the streets of the historical centre (*centro histórico*), which aims to privatise public space through an exclusionary, neoliberal urban development programme. Yet traders have voiced concerns over the lack of state recognition of the impacts of displacement policies on their families. According to one trader, 'what the [*centro histórico*] doesn't understand is that we are not merely 48 street vendors who will be dislocated. We are 48 families' (Véliz and O'Neill, 2011: 95). Traders also speak of the important role of markets as places to meet friends, be with family and socialise with community members; some traders have worked in the city for most of their lives and have deep social ties to the streets. In Mexico City, Crossa (2009) likens the streets to 'home'. As one trader states,

> here I feel at home. Like with my family. I mean, among us we really help each other. There is more communication here among us than in my own house where I just eat and sleep. This is like one big family. We basically live here.
>
> Ibid.: 56

Yet, despite this, in some cities punitive urban policies continue to forcibly remove street traders from city streets.

Mete *et al.* (2013) offer some hope that the socio-cultural battleground is not entirely lost. In their Mexican study they claim that the 'rootedness' of street trading in local cultural heritage has enabled them to remain on the streets, despite government efforts to remove them from public spaces.

Street trading and Latin America's post-revanchist turn

Studies of street trader displacement have rarely been longitudinal, but have tended to document the displacement process and its *immediate* impacts on traders. Therefore, there is limited evidence as to what happens when 'the class politics embodied in revanchism plays itself out' (Murphy, 2009: 311). It could be expected that displacement policies would persist, for example, through policing public space, but in a rare longitudinal study in Peru, Mackie *et al.* (2014) documented a shift in policy discourse and practice, arguing that previous revanchist policies targeted at street traders have given way to a more supportive and tolerant post-revanchist era. The lack of significant evidence of post-revanchism may be due to the lack of longitudinal research, but may equally reflect the continued dominance of revanchist urbanism which, as this chapter demonstrates, still persists with vigour.

The next section of the chapter aims to explore the nature of post-revanchist policy-making and its impacts on public space, again using Mackie *et al.*'s (ibid.) battlegrounds framework. The discussion draws upon the limited selection of case studies where more tolerant, supportive policies have been documented, including Peru and Colombia.

Spatial tolerance

The fight over space in the city lies at the heart of revanchist urban policy-making, so perhaps the most notable indication of a post-revanchist policy is the renewed tolerance of street traders in the city centre. In Cusco and parts of Colombia, where traders have historically been removed from the city streets in large numbers, policy and practice is emerging which permits trade on the street and tolerates alternative trading sites in the city centre.

Mackie *et al.* (2014) found that in 2011 the municipal authorities in Cusco were developing regulations that would permit street traders to sell on the streets. The policy was in its infancy in Cusco and had not yet been implemented, but marked a clear shift from earlier policies of displacement. According to Mackie *et al.* (ibid.), policy in Cusco was relatively vague but intended to limit *all* traders to a small number of trading days on the street. Municipal authorities are clearly grappling with the difficulties of a changing practice which permits some trading on the streets, by imposing caveats on when people may trade, to avoid a return to a situation where street traders occupy many of the city centre thoroughfares.

Courtyard markets, defined as developments within the courtyards of old buildings where groups of traders rent space from the property owner, emerged as an unplanned and street trader-led response to street trader relocations in Cusco (Bromley and Mackie, 2009). Mackie *et al.* (2014) document how these alternative spaces, located in the city centre, were once the subject of municipal government closure notices, but, in 2011, they had become an extension of the streets, with 17 new markets opening between 2006 and 2011. Mackie *et al.* (ibid.) also report how traders in these markets were now able to sell without fear of police harassment. It is clear that in Cusco there has been a shift from staunch policies of displacement to a collection of context-specific policies which tolerate, to varying degrees, the presence of street traders in the city centre.

In Colombia, the national constitution has been pivotal in creating a more tolerant approach to the presence of street traders in public spaces. According to Meneses-Reyes and Caballero-Juarez (2014) street traders have a right to sell on the streets on the assumption that they are economically 'disfavoured' and therefore removing them from the streets would infringe their right to work. Under this system, municipal authorities can revoke licences to trade in particular places only if traders can exercise their right to work elsewhere. Despite this apparently tolerant and supportive approach, Galvis (2014) documents how itinerant street vendors remain excluded in practice, arguing that deep-seated forces of displacement embedded within the nation's social, racial and class structure continue to exclude traders from working in the city centre.

Political tolerance

The political battle to have one's voice heard is a longstanding struggle for most marginal groups, and street traders are no different. Earlier in the chapter it was demonstrated that an era of revanchist policy-making and lack of voice for street traders resulted in exclusion from city centre public space and detrimental economic and social changes. The post-revanchist turn is characterised by a shift in the degree of participation and power of street traders, a shift that has been borne out to some extent in Cusco.

The power shift in Cusco was achieved through mayoral elections. Mackie *et al.* (2014) document how the significant street trader populations have been courted for their votes by offering promises of more liberal regulation of trade. Indeed, the election of Mayor Sequinos in 2007 brought about a considerable increase in street traders working on the streets. While such political changes are evidently post-revanchist, the sustainability of a power shift dependent on a single mayor is relatively weak. Indeed, there was some reversal of Sequinos's policies in Cusco after she left office.

Economic tolerance

In the few cities where there has been a shift from punitive policing in public space (revanchism) to more supportive practice (post-revanchism) there appear to be changing perspectives on the economic needs of street traders. According to Mackie *et al.* (ibid.), changes in Cusco allowing traders to return to the streets, at least for certain periods, have been largely driven by a recognition that many traders need to occupy public spaces in order to survive. Earlier relocation policies often led to spatial resistance, hence city authorities now recognise that, while there is an economic need and an evident return to the streets, the activity should be permitted and regulated. However, in Cusco policies fall short of truly recognising street trading as a desirable element of the city economy, evidenced by the rhetoric of capacitation and formalisation which accompanies policies targeted at street traders (ibid.).

As in Cusco, it is the economic needs and rights of traders that have led to more tolerant policies about the use of public space in Colombia. Meneses-Reyes and Caballero-Juarez (2014) describe how constitutional courts in Colombia have, in some instances, determined that street traders should be allowed to occupy public space in order not to infringe upon their right to work. Despite the more tolerant nature of policies in Cusco and Colombia, neither approaches take steps to recognise the value of the informal economy. They fail to address Middleton's concerns that street traders have something to contribute and their involvement in local economies should be supported (2003).

Social and cultural tolerance

When revanchist urbanism takes hold, the cultures and practices of street traders are often dismissed to make way for those of more dominant groups. While the post-revanchist turn has been characterised by gains in spatial, political and economic battles, to date there is limited evidence of any significant gains in the socio-cultural battleground. For example, capacity-building policies in Cusco, while largely couched by policy-makers as a form of support, take no account of the possibility that street traders may not wish to formalise, or that consumers may prefer the flexible and informal nature of street trade. In Cusco, Mackie *et al.* (2014) argued that capacity-building was just one tool used to formalise street trade; they also brought in a new class of traders in the hope that street traders might replicate their *good* practices.

Conclusions

The detail of regulation and practice of urban management and policing combine to provide a powerful system of control over public space, with profound impacts for street traders. This chapter has explored how different policy approaches can result in very different outcomes for traders. Since the emergence of zero-tolerance policing in Giuliani's New York, revanchist urbanism has been the dominant urban policy paradigm across Latin America. Consequently, street traders, like many other marginal groups, have been targeted for removal from city centres across the region. Using Mackie *et al.*'s (ibid.) battlegrounds framework, we have illustrated how revanchist urbanism persists with vigour today in parts of Brazil, Colombia, Ecuador, Guatemala, Mexico and Peru. It is a paradigm that is forcing some street traders to undesirable trading locations where they face poverty and resort to practices such as transnational migration. Their vulnerability is compounded by the disruptive effects of displacement on supportive social networks and trader associations.

In contrast, Latin America is also the birthplace of a potentially more tolerant and supportive post-revanchist turn. Although the punitive paradigm is still evident, we argue that in some cities supportive policy and practice is emerging alongside existing control. This emerging softer approach may help foster a more inclusive city. We hope that policy-makers across the globe might reflect on the lessons which are emerging from these case studies. First, no single social or economic group should have outright dominance over urban space; rather everyone should have access to it. Efforts to allow street traders to return to city centres in Cusco and Colombia are a step in the right direction. Second, planning processes should be participatory and inclusive of all marginal groups. While mayoral elections offer some opportunity for voices to be heard, ongoing participation is necessary for sustainable solutions.

Third, the post-revanchist turn has seen greater recognition of the economic needs of street traders, suggesting that policies must be tested for their impacts on those who are already marginalised and are often very poor. In Cusco and Colombia, the economic needs of street traders have been a key driver behind the decision to permit traders back onto the streets. Failure to challenge policies with detrimental economic impacts will only result in poverty, traders returning to the streets, or actions such as international migration. The battle in which the greatest fight remains is the battle for social and cultural equality. City governments appear to have much to do before equal treatment of different cultures is achieved in practice. Until then, activities such as street trading will continue to be unfairly treated and the traders detrimentally affected.

Cities across the globe are grappling with the need to balance punitive and supportive approaches to dealing with marginal urban groups such as street traders, a battle being fought in public space. Lessons from our review of policy and practice in revanchist and post-revanchist cities will therefore have wider relevance for urban policy-makers, both in their efforts with street traders and with marginal groups more generally. Our hope is that the post-revanchist turn takes hold as rapidly and as widely as its revanchist predecessor to provide a new, more supportive regime for managing urban space.

References

Becker, A. and Müller, M.-M. (2013) The securitization of urban space and the 'rescue' of downtown Mexico City: vision and practice, *Latin American Perspectives*, 40(2): 77–94.

Bowling, B. (1999) The rise and fall of New York Murder. Zero tolerance or crack's decline? *British Journal of Criminology*, 39(4): 531–54.

Bratton, W. and Andrews, W. (2001) Driving out the crime wave: the police methods that worked in New York City can work in Latin America, *Time Magazine*, 23 July.

Bromley, R.D.F. and Mackie, P.K. (2009) Displacement and the new spaces for informal trade in the Latin American city centre, *Urban Studies*, 46(7): 1485–506.

Brown, A. (2006) *Contested Space: Street Trading, Public Space, and Livelihoods in Developing Cities*, Urban Management Series, Rugby: ITDG.

Carrieri, A.-de-P. and Murta, I.B.D. (2011) Cleaning up the city: a study on the removal of street vendors from downtown Belo Horizonte, Brazil, *Canadian Journal of Administrative Sciences*, 28(2): 217–25.

Crossa, V. (2009) Resisting the entrepreneurial city: street vendors' struggles in Mexico City's Historic Center, *International Journal of Urban and Regional Research*, 33(1): 43–63.

Crossa, V. (2013) Play for protest, protest for play: artisan and vendors' resistance to displacement in Mexico City, *Antipode*, 45(4): 826–43.

Davis, D. (2013) Zero tolerance policing, stealth real estate development, and the transformation of public space: evidence from Mexico City, *Latin American Perspectives*, 40(2): 53–76.

DeVerteuil, G. (2014) Does the punitive need the supportive? A sympathetic critique of current grammars of urban injustice, *Antipode*, 46(4): 874–93.

Donovan, M.G. (2008) Informal cities and the contestation of public space: the case of Bogota's street vendors, 1988–2003, *Urban Studies*, 45(1): 29–51.

Galvis, J.P. (2014) Remaking equality: community governance and the politics of exclusion in Bogota's public spaces, *International Journal of Urban and Regional Research*, 38(4): 1458–75.

Harcourt, B.E. and Ludwig, J. (2006) Broken windows: new evidence from New York City and a five-city social experiment, *University of Chicago Law Review*, 73(1): 271–320.

Hunt, S. (2009) Citizenship's place: the state's creation of public space and street vendor's culture of informality in Bogota, Colombia, *Environment and Planning D: Society and Space*, 27(2): 331–51.

Little, W.E. (2005) Getting organised: political and economic dilemmas for Maya handicraft vendors, *Latin American Perspectives*, 32(5): 80–100.

Mackie, P.K., Bromley, R.D.F. and Brown, A. (2014) Informal traders and the battlegrounds of revanchism in Cusco, Peru, *International Journal of Urban and Regional Research*, 38(5): 1884–903.

Meneses-Reyes, R. and Caballero-Juarez, J.A. (2014) The right to work on the street: public space and constitutional rights, *Planning Theory*, 13(4): 370–86.

Mete, S., Tomaino, L. and Vecchio, G. (2013) Tianguis shaping ciudad. Informal street vending as a decisive element for economy, society and culture in Mexico, *Planum: The Journal of Urbanism*, 26(1): 1–13.

Middleton, A. (2003) Informal traders and planners in the regeneration of historic city centres: the case of Quito, Ecuador, *Progress in Planning*, 59(2): 71–123.

Mountz, A. and Curran, W. (2009) Policing in drag: Giuliani goes global with the illusion of control, *Geoforum*, 40(6): 1033–40.

Murphy, S. (2009) 'Compassionate' strategies of managing homelessness: post-revanchist geographies in San Francisco, *Antipode*, 41(2): 305–25.

Musoni, F. (2010) Operation Murambatsvina and the politics of street vendors in Zimbabwe, *Journal of Southern Africa Studies*, 36(2): 301–17.

Peña, S. (1999) Informal markets: street vendors in Mexico City, *Habitat International*, 23(3): 363–72.

Smith, N. (1996) *The New Urban Frontier: Gentrification and the Revanchist City*, New York: Routledge.

Smith, N. (1998) Giuliani time: the revanchist 1990s, *Social Text*, 57: 1–20.

Swanson, K. (2007) Revanchist urbanism heads south: the regulation of indigenous beggars and street vendors in Ecuador, *Antipode*, 39(4): 708–28.

Swanson, K. (2010) *Begging as a Path to Progress: Indigenous Women and Children and the Struggle for Ecuador's Urban Spaces*, Athens: University of Georgia Press.

Turner, S. and Schoenberger, L. (2012) Street vendor livelihoods and everyday politics in Hanoi, Vietnam: the seeds of a diverse economy?, *Urban Studies*, 49(5): 1027–44.

Véliz, R.J. and O'Neill, K.L. (2011) The displacement of street vendors in Guatemala City, in O'Neill, K.L. and Thomas, K. (eds), *Securing the City: Neoliberalism, Space and Insecurity in Postwar Guatemala*, Durham, NC: Duke University Press: 83–102.

Wacquant, L. (2003) Toward a dictatorship over the poor? Notes on the penalization of poverty in Brazil, *Punishment and Society*, 5(2): 197–205.

Wilson, J. and Kelling, G. (1982) Broken windows, *The Atlantic Monthly*, March: 1–10.

6 Claiming the streets

Reframing property rights for the urban informal economy

Alison Brown

Summary

This chapter looks at the concept of law and urban development with a focus on property rights and land, exploring the potential of a new concept – that of collective rights in the public domain – to underpin a more equitable approach to the management of public space and challenge inappropriate regulation that criminalises the lives of the poor. The research draws on fieldwork in Dakar and Dar es Salaam, undertaken in 2010–12 as part of comparative research funded under the joint Economic and Social Research Council (ESRC)/Department for International Development (DFID) poverty reduction programme on law, rights and regulation and street trade.

Introduction

For many years, the provision of secure tenure has been a key strategy in the approach to upgrading informal settlements. The focus has been on land for housing, rather than for small-scale economies, and the public domain has been largely excluded from these debates, as being somehow a different form of urban space. Yet, for many of the urban poor with limited space in the home, external space – often the street – plays a multitude of roles, as a place for socialising, play and ceremony, and also a crucial but highly contested place of work. Yet this so-called 'public space' is, in fact, far from 'public' in terms of access and rights of use, and the continuing evictions of poor traders from city streets is a global scandal which has received far too little attention.

This chapter looks at the concept of law and urban development with a focus on property rights and land, exploring the potential of a new concept – that of collective rights in the public domain – to underpin a more equitable approach to the management of public space and challenge inappropriate regulation that criminalises the lives of the poor. The research draws on fieldwork in Dakar and Dar es Salaam, undertaken in 2010–12 as part of the comparative research programme on law, rights and regulation and street trade funded under the DFID/ESRC Joint Fund for Poverty Alleviation.

Figure 6.1 Trading rights in public space are complex and negotiated: Grand Marché, Lomé, Togo, and wall space, Cairo, Egypt.

The chapter has five sections. After the introduction, section two examines the literature on property rights in public space, followed by a brief critique of the treatment of property rights in the report of the Commission for Legal Empowerment of the Poor (CLEP). Sections three and four draw on fieldwork in Sénégal and Tanzania. The fieldwork included around 60 key informant interviews and interviews with 143 traders in six trading locations in Dakar, and 173 interviews in five trading locations in Tanzania. The final section argues for a rethinking of rights in public space.

Property rights and public space

The concept of 'public space' is widely debated in different disciplines covering issues of design, privatisation, deliberative democracy and the political economy of space (e.g. Low and Smith, 2005; Webster and Lai, 2003; Harvey, 2003). However, these discussions often fail to consider urban livelihoods and the need for secure rights to use public space (Brown, 2006). In this respect the CLEP report (UNDP, 2008) was a welcome departure in considering property rights as one of the four pillars of legal empowerment essential to business rights for the poor. This section briefly introduces concepts of property rights and collective tenure.

A property-rights regime reflects the social organisation of communities and determines to a reasonable degree the activities which people can undertake within in a specific domain. Rights are expressed through the 'laws, customs and mores of a society' (Demsetz, 1967) and depend on the existence of an enforceable set of rules. Property rights are usually considered as *bundles of entitlements* which may include *access, beneficial use, exclusion* and *transfer*, with a scale of rights extending from full *private ownership* (excludable and transferable) to *communal ownership* with full public access and no rights of transfer to any other party. *State ownership* is sometimes a third category defined as implying rights to determine the use of property within politically defined processes (ibid.).

In much of pre-colonial sub-Saharan Africa the dominant system of land entitlements depended on communal rights, often referred to as *customary tenure*. Land was assumed to be a sacred resource where use rights are granted on the basis of need, with responsibilities for allocation of use rights and protection of the resource vested in the king or tribal elders (Delville, 2000; Hesseling, 2009; Alden, 2002). Under colonial administration, decrees established rights that mirrored property legislation in Europe, mainly for annexed settler land and towns. Attempts at land reform largely failed, as the indigenous peoples saw no reason to conform to imposed legislation which was considered both 'useless and dangerous' (Hesseling, 2009).

Governments and the international community considered private property rights as essential to economic development, and saw communal tenure as archaic and inefficient. In the African context, colonial legislation introducing private registered ownership did not take account of traditional customs, where the existence of different forms of land tenure is common (Delville, 2000). In a study of rural Tanzania, Odgaard (2003) found six different forms of *de facto* tenure: customary rights accorded to indigenous groups; customary rights accorded to trustworthy in-migrants; rights allocated by the village authorities; borrowed or rented land rights; rights obtained through commercial transaction often by urban investors; and open access to non-allocated land.

Ostrom and Hess distinguish three types of communal rights (2010: 56–9):

- *common property regimes*: closed regimes where members of a defined group have access to a resource with a legal right to exclude non-members;
- *open access regimes*: regimes that allow anyone free access. However, unless there are accepted rules that govern access no-one has an incentive to invest in protecting the resource which leads to over-use, a problem famously described in Hardin's 'tragedy of the commons' (Hardin 1968);
- *common-pool regimes*: a hybrid between closed and open regimes, where there is common ownership but it is costly or impractical to exclude people either physically or through regulation. Here, use by one person detracts from the use by others, and the resource is liable to over-use and congestion unless limits are devised.

Public space used for street trading is probably best described as a common-pool resource because access is difficult to control, but self-regulation is difficult to achieve because of the shifting populations of city centres. Webster (2007) uses a property-rights analysis to describe how, in high-density cities, urban public spaces suffer competition for space, leading to a dynamic of urban change. He argues that urban public spaces are collectively consumed goods – at low levels of consumption they are non-rivalous and non-excludable – ie one person's use does not deter another's – but over time spaces become over-used and congested, leading to demands to restrict their use, effectively reassigning property rights. Webster (ibid.) concludes that only by restricting the use can the resource be protected and maintained, suggesting that, for hawkers

Figure 6.2 Public with *de facto* rights: trading space in Dakar and Dar es Salaam

in a busy street, congestion could be managed through pricing, limiting the range of goods sold, defining pitches, or creating Singapore-style hawkers' malls, though an outright ban would probably fail.

In the African context, local land systems are often flexible and evolving, so customary law can be seen as 'procedural' rather than codified (Delville, 2000, 2006). In Francophone regions property management was often centralised, land was nationalised, but use rights were recognised in village land, as in Sénégal. In Anglophone Africa the system of indirect rule left more room for control by traditional authorities, but the continued existence of dual and often overlapping systems causes uncertainty about rights and leads to ongoing land disputes, often manipulated by local elites. Delville (2000) proposes two solutions:

- *registration*: harmonising legal systems under formal law, creating a single structure of property rights, based on cadastral surveys, or a partial system with legal categories that match local custom – for example, recognising customary title, or acknowledging different types of use rights over the same area;
- *common heritage*: recognising collectively apportioned land distributed through transparent rules and arbitration. This approach allows state and communal systems to coexist, with an emphasis on the process of establishing rights.

Thus adopting a property-rights analysis of the 'street as a place of work' suggests *two key additions to an understanding of rights*:

1) Urban public space is part of a continuum of urban land and should be encompassed within any property-rights regime, recognising that *de facto* rights in urban public space exist and are crucial to the livelihoods of the poor. This is best described as a *common-pool resource* that needs to be managed

for shared benefits to accrue, with defined 'bundles' of rights – for example, allowing access and beneficial use.

2) Although street vending and other work in public space involves at least temporary appropriation of space for private gain, its collective use contributes significantly to urban employment and urban service provision, thus both to poverty reduction and urban management.

Thus, rights in urban public space cannot be identified as private rights as they are neither excludable nor transferrable, and are often communally managed and maintained. The challenge lies in managing the competing demands on this communal resource.

Legal empowerment and livelihoods

CLEP published its report, *Making the Law Work for Everyone* in 2008 under the aegis of UNDP (UNDP, 2008). The report argues that four billion people around the world are excluded from the rule of law, and that, drawing on human rights principles, access to legal processes is a crucial component of poverty reduction. The Commission identified four pillars of legal empowerment: access to justice and the rule of law; property rights; labour rights; and business rights. This chapter examines the second pillar – property rights – and their application to business rights.

The CLEP report draws heavily on the views of co-chair Hernando de Soto. His main thesis is that formal property rights are crucial to effective functioning of markets and economic growth, and that lack of legally protected transferrable property rights denies the poor access to credit and limits their potential for capital accumulation (de Soto, 1989, 2000). Critics of de Soto's approach have argued that the requirement for formal titles discriminates against those without occupancy rights or with collective title, and excludes those who cannot pay for formalisation, and that the economic benefits of secure title are over-stated (Davis, 2006; Gilbert, 2002).

Four dimensions of property rights in the CLEP report are seen as important: a system of rules that defines bundles of rights and obligations; a system of governance; a functioning land market; and an instrument of social policy (UNDP, 2008: 66). Certain groups are seen as particularly vulnerable to insecure property rights, including women, indigenous people and slum dwellers. Formal property systems often evolve to benefit political and social elites who resist change, including large landowners, lawyers, state officials profiting from bribes and entrepreneurs avoiding regulation (ibid.: 80).

Interestingly, the report does deal with collective rights, moving away from de Soto's original emphasis on individual rights, suggesting that pooling assets in transparent co-ownership structures and legally recognised common rights can give social and economic leverage. The report argues that the state should not be the default owner of land or property, and that zoning and city planning are particular instruments of exclusion of the poor (ibid.: 75–97). Simplified property registration can confer legal rights of use or occupancy as alternatives to full title (Durand-Lasserve and Royston, 2002).

The report deals tangentially with rights in public space, suggesting that basic commercial rights should include a right to work and right to a work space (including public land and private residences), and to appropriate infrastructure (UNDP, 2008: 201). This requires legislation (municipal bylaws) that allow street traders to operate in public spaces, which define use rights to urban public land and appropriate zoning regulations governing the conditions under which informal businesses can operate in central business districts and other areas (ibid.: 221). Unfortunately, these ideas are not elaborated, and are somewhat contrary to the approach in an earlier chapter, which suggests that rights for street traders can best be achieved in 'delineated hawker zones' (ibid.: 101).

The over-riding assumption of the CLEP report is that formalisation is the only route forward, and there are many gaps in the discussion which are not fully explored, for example: what bundles of property rights might be defined; the merits of different types of collective or communal ownership; the challenges in ensuring benign and transparent state action on property rights; and different approaches to land rights needed in urban and rural contexts. A crucial problem in urban areas is that very high proportions of the urban poor are renting accommodation whose rights are often ignored in the process of securing tenure.

A political tightrope: Sénégal

The next two sections of this chapter examine findings from studies in Dakar and Tanzania, and the extent of informal claim that exists in public space.

Sénégal is one of the most stable countries in West Africa and, in the last two decades, has witnessed peaceful elections and transfer of political power. Sénégal's population is around 14.7 million, with a GNI of US$1,050 per capita (2014 estimates); 46.7% of the population is below the national poverty line; the level of urbanisation is 43%, with 39% of the urban population living in slums (2014 estimates) (World Bank, 2016a). The majority of the population is Muslim; Wolof is widely spoken in northern coastal regions, with the other main languages being Perle (Pular) and Serer. The capital city of Dakar had an estimated population of 3.4 million in 2014.

Parallel systems of land rights

Land reform in reform in Sénégal took place shortly after independence, and *Loi no. 64-46 relative au Domaine National* (1964 Land Law) was widely heralded as an innovative attempt to harmonise formal land ownership with customary rights; but, 50 years on, implementation is still partial (Dankoco, 2011; Benkahla and Seck, 2011; Boye, 1978). Like many countries in the region, Sénégal has retained a dual system of land tenure, drawn from the pre-colonial customary traditions and the French Civil Code governing land (Hesseling, 2009).

Colonial era legislation still defines the legal structure of land administration in Sénégal. Regulations from the early 1900s sought to facilitate registration of customary rights, but were unsuccessful because they granted individual but not collective rights – by far the most important in African culture at the time. Decrees in 1955 and 1956 sought to transform both individual and collective rights into full and individual property rights (ibid.).

The problem of parallel property regimes was recognised in a study on land affairs, published shortly before independence, which proposed a constitutional guarantee of property rights. The resulting 1964 Land Law (amended by Law 76-66) reflected the ideology of Senegalese socialism and the desire to redress the colonial legacy by enabling all citizens to access land (Benkahla and Seck, 2011). The Act decreed that all land belongs to the state, with the exception of land with formal registered title, and the law now recognises three broad categories of land rights: *state domain*; *private domain*, regulated by the 1932 decree and registered in the *Livre Foncier* (Land Book); and the new *national domain*, which covers all non-registered land – divided into urban zones, classified zones (protected areas), *zones de terroirs* (village land) and pioneer zones (undefined areas) (Hesseling, 2009).

Three categories of land holdings were set out. The *titre foncier* is a registered title which enables land to be bought and sold; the *permis d'occuper* is an occupancy right which can be revoked at any time by the state; and the *lease* retains land ownership with the state, but allows use of land on a rental basis (Dankoco *et al.*, 2010). Village land is allocated to those who demonstrate productive use (*mise en valeur*) administered by the *conseils ruraux* created in 1972, although these are still struggling to implement the legislation (Hesseling, 2009).

Thus, for a long time, dual systems of land rights have existed in Sénégal. There are also thriving informal markets in urban land, particularly in urban fringe areas. Near Dakar land is often purchased by city dwellers from village chiefs or traditional owners. Although sales of national domain land are illegal, the agreement is often ratified by the rural council in a document that gives a *droit d'occupation* (ibid.). In urban areas, the legislation presumes that private domain land pertains, but the existence of parallel markets is still common.

Much of the literature on land rights in Sénégal focuses rural land, but it is evident that over a century of attempts to introduce a unified system of land legislation have largely failed. Rights are an expression of social organisation and custom, and as rural and urban communities share similar philosophies and ideals, attitudes to urban land in Sénégal are therefore likely to be both pluralist and pragmatic. While discussion of land tenure in urban areas generally focuses on housing land, work in other countries of sub-Saharan Africa has shown that bundles of rights also pertain to urban public space (Brown, 2006: 202) – indeed, the distinction between public and private space is itself a product of imported European norms.

The political economy of trading space in Dakar

For over two decades, street trade in Dakar, like the wider business environment, has been a highly political. In the early years after independence, European (mainly French) capital dominated the formal sector in industry, commerce and banking, with smaller businesses run by companies of Lebanese origin (Thioub *et al.*, 1998). Influential African businesses gradually won control over lucrative imports, and the Islamic brotherhoods, which grew during the nineteenth century to control the peanut trade, took advantage of contraband routes through the Gambia (Lyons and Snoxell, 2005; Brown *et al.*, 2010). This paved the way for the emergence of a flourishing of informal economy, largely tolerated by the government as a way of providing jobs and cheap goods during the high inflation and economic recession of the 1970s and 1980s (Thioub *et al.*, 1998).

Between 1986 and 1990 structural adjustment and neoliberal economic reforms led to privatisation of state industries and fights to control the lucrative import trade (ibid.). In 1989 attempts by the government to increase customs duties and sales tax (TVA) were foiled when businesses protested by closing shops and offices, paralysing the central business district (CBD). In 1990 the *Union National des Commerçants et Industriels du Sénégal* (UNACOIS) was formed, a broad-based organisation that fought to protect informal business interests. Operating through strikes and direct action, particularly in 1993 and 1996, UNACOIS successfully fought off import tax increase, and attempts to apply TVA to street traders. By 1998 UNACOIS claimed a membership of 70,000, including many small-scale traders and hawkers (ibid.).

The 2000 election of President Abouldaye Wade heralded a change in attitudes towards street traders. The Wade campaign courted traders and, after this, merchants gained confidence, stopping the new president's cavalcade on the way from the airport; he agreed that traders should not be evicted. This was a significant event; many hawkers built kiosks along the roads and an influx of new traders occurred.

In November 2007, the President issued an instruction to the Governor of Dakar region to tidy the streets. This was interpreted as an instruction to evict. UNACOIS was informed of the move, but the information did not reach many traders on the ground. Clearances took place on 21 November, resulting in three days of riots, effectively shutting down the city centre. With perhaps a *volte face*, the president dismissed the governor and announced negotiations with the vendors, provided that they were organised. A consultative commission was set up with the *Ministère du commerce*, and the *Mairie de Dakar* set up a weekly consultation group with traders' representatives (key informant interviews).

In the wake of the 2007 riots, one of the more radical trader factions tried to join UNACOIS, which decided, however, that the group would been seen as an opposition force by the government. Several new traders' organisations were then formed in 2009, including *Synergie des marchands dits ambulants pour le développement* (SYMAD), comprising 12 associations with a combined membership around 7,000, and *Fédération des Associations des Marchands Ambulants et Tabliers du Sénégal* (FAMATS), with 15 associations and over

5,400 members (key informant interviews). SYMAD proposed that street trading be legalised and managed at street level by vendors.

Meanwhile, in a shift from its previous militancy, UNACOIS showed a distinct reluctance to support street vendors, a sector which it described as essentially illegal, in 2010 reportedly making a statement of support for the evictions to the *Mairie* (key informant interviews). UNACOIS sought to restrict the influx of traders during public holidays such as the Muslim festival of Tabaski, but also supported a three-day strike against an increase in the daily toll – so perhaps the attitude is pragmatic.

In 2007, the *Ministère du commerce* took a census of street traders in Dakar, recording around 8,000 traders, thought to have increased to 15,000 by 2010. Traders' groups disputed this estimate, which appeared to be an under-count if the membership figures for SYMAD and FAMATS were correct. The census reported that 91% of traders were men, 82% were under the age of 30 and 80% were funded by credit from suppliers. An international symposium on formalising street trade was arranged with the International Labour Office (ILO/*Bureau International du Travail, BIT*) and Sénégalese government in 2008.

At national level, there appears to be a general willingness to resolve the problems of street traders. Sections 8 and 25 of the national constitution establish the right to work:

s.8: The Republic of Sénégal guarantees all citizens fundamental freedoms, economic, social and collective rights. These rights and freedoms include. . . the right to work

s.25: Everyone has the right to work and the right to claim employment. No-one shall be discriminated against in their work due to their ethnicity, sex, opinions, choices or political beliefs.

But these are far from being implemented, and traders are vulnerable to shifts in policy.

The *Mairie de Dakar* has purchased a site in rue Félix Eboué, costing FCFA 2 billion, for 1,500–2,000 vendors, to clear the downtown streets. The market is now complete, but there are concerns that the site is liable to flooding and that kiosks will be too expensive; traders have been reluctant to move in. New sites are also being sought for those displaced in 2007 in Parcelles Assainies, Thiaroye and Mbeubeuss. In autumn 2010 the President declared construction of a new market at Pètersen, and in February 2011 he announced the creation of a new agency charged with addressing the issue of the *marchands ambulants*, although this was disbanded by the new government of 2012.

Tentative gains?

High-level political support seems to have secured a degree of stability for traders, but ongoing harassment from low-level bureaucrats is rife, and poor traders remain vulnerable. Traders interviewed considered they had rights to space. However, these rights were contested and harassment and evictions remained a critical problem for the traders. Of the 143 interviews, 90% had experienced dimensions of insecurity and 34% had experience of eviction. An experience typical of many is as follows:

> *Un jour, lors d'une opération de déguerpissement, je discutais avec la police. Ils m'ont enfermé un mois de prison et j'ai payé 50,000 FCFA pour sortir de la prison. En ce moment j'empruntais de l'argent pour nourrir ma famille.* (One day during an eviction operation, I argued with the police. They locked me up for one month in prison, and I paid 50,000 FCFA to get out of prison. At that time I was borrowing money to feed my family.)
>
> Male trader, selling underwear, rue Blaise Diagne

Thus, over a period of 20 years, traders have claimed rights through a process of direct action which creates space for political negotiation. This has enabled traders to develop a degree of organisational capacity around area-based or product associations, thus breaking the traditional clientelist relations between petty traders and the large-scale wholesalers. There is legal acceptance of plural land rights, though not applied to public space, and potential for communal management of trading rights, but turf wars between traders' associations over space and influence with government officials seem likely to undermine some of the gains achieved. Meanwhile, the illegal character of street trade means that any gains will be tentative.

Continuing struggle: Dar es Salaam

Like Sénégal, Tanzania also has a dual land code and a socialist tradition, but fluctuating policy towards street trading. Shortly after independence in 1961, the 1967 the Arusha Declaration heralded in the uniquely Tanzania policy of *Ujamaa* (socialism and self-reliance) and one-party rule, but wars and a weak economy led to a period of structural adjustment. Multiparty politics were reintroduced in 1992, but, on the mainland, power remains in the hands of the long-ascendant Chama Cha Mapinduzi (CCM) party, although opposition parties now rule in Zanzibar. World Bank data (World Bank, 2016b) sets the country's population at 52 million, with a GNI of US$920 per capita (2014 estimates); some 28% of the population was below the national poverty line. Urbanisation levels are estimated at 31%, with 50.7% of the urban population living in slums (2014 estimates). Dar es Salaam had an estimated population of 4.8 million in 2014.

Land and property rights in Tanzania

Land regulation in Tanzania, as in Sénégal, has undergone more than a century of change in an attempt to reconcile the colonially imposed land rights with the norms of customary tenure. Under colonial rule, all land was decreed as crown land except where individuals had documented ownership or traditional rights (Shivji, 1998; Olenasha, 2004). After independence in 1961, the 1965 Constitution affirmed government control over land; freehold title was abolished and converted to a government lease (Olenasha, 2004), an effective nationalisation of land.

The 1990s was a period of major reform in the conceptualisation of land and local government administration. In 1991, the Shivji Commission on Land Matters reviewed land laws and their administration, and a National Land Policy was published in 1995 (Shivji, 1998; Kombe, 2000; Olenasha, 2004; Mallya, 2005; Hillbom, 2011). Continuing the principle of state ownership of land, the policy recognised that 'all Land in Tanzania is Public Land vested in the President as trustee on behalf of citizens', and sought to give legal recognition to long-standing occupation; reflect the principles of sustainable development; and pay fair compensation for revocation of occupancy rights (URT, 1997: s.4.1.1). The policy led to the publication of two lengthy pieces of legislation, the 500-page Land Act (No 4) 1999, and the 280-page Village Land Act (No 5) 1999 (Olenasha, 2004).

For non-village land, the Land Act 4/1999 (URT, 1999a) establishes a 'right of occupancy' of up to 99 years, where permission is required for change of use of development (s.35-1) and 'derivative rights' or annual residential licences (s.19-1) for to those who had occupied urban or peri-urban land for up to three years before commencement of the Act (URT 1999a, s.23-2). The Village Land Act 5/1999 specifies three types of land: village land, general land and reserve land, establishing a 'certificate of customary occupancy' which can only be granted to Tanzanian citizens on customary land (Mallya, 2005; Olenasha, 2004).

State ownership of land remains a central plank of policy in Tanzania, but with recent moves to introduce a functioning commercial land market. The Land Act 2004 introduced amendments promoted by the banking sector, for the first time giving formal value to undeveloped land, enabling vacant land to be bought and sold and facilitating the use of land as collateral (URT, 2004; Olenasha, 2004). Olenasha (2004) criticised the reforms as 'blindly echoing' the models of de Soto, which strips land of its cultural and spiritual values and could lead to wide-spread dispossession and destitution.

In practice, informal land markets have existed in unplanned urban settlements for a long time. A 1994 study in Manzese found that two-thirds of 99 respondents of home 'owners' had bought their land or house (Kironde, 2000); similar find-ings emerged from another study of 61 households in the unplanned areas of Hanna Nasif and Mogo (Kombe, 2000). Nevertheless, despite the comprehensive reforms in the 1999 Acts, implementation was slow. So, in 2004, de Soto was invited to Tanzania and his consultancy, the Institute of Liberty and Democracy (ILD), then advised the government on property and business reform under the Property and Business Formalisation Programme known by its Kiswahili acro-nym of MKURABITA (*Mpango wa Kurasimisha Rasilimali na Biashara za Wanyonge Tanzania*).

MKURABITA's aim is to empower the poor by increasing access to property and business opportunities to underpin the expansion of Tanzania's market economy (URT, 2008a). Although a flagship of a previous administration, the recommendations are being quietly pursued and include measures to amend the 1999 Acts, to strengthen land rights, streamline registrations, simplify adjudication, establish a unified land registry and facilitate access to mortgages (URT, 2008b). Village land remains under customary title, but in urban areas the thrust of reform is to make a swift transition to a market economy in land.

MKURABITA is seeking to streamline the process of survey and titling. A pilot project was undertaken in Hanna Nasif, an unplanned settlement with a population of 32,000, to register 1,131 informal plots. The project was run by the NGO the Women's Advancement Trust (now WAT-Human Settlements Trust, WAT-HST) and demonstrated that community consultation combined with a block registration process could achieve a low-cost registration process (key informant interviews). Yet, despite the recognition of dual land rights in the 1999 Acts, government initiatives seem to promote a unified system.

The battle for the streets

Use of town planning powers in Tanzania has also been a tool of dispossession, and since 2003 legislation has progressively marginalized traders (Brown *et al.*, 2010). In 2003 the *nguvukazi* (peddler's licence) was cancelled, effectively making street trade illegal. Growing donor emphasis on increasing tax revenue, and the adoption of western ideals of city planning, compounded the pressures for formalisation.

The larger street trader associations in Tanzania have been surprisingly weak. During the mid-1990s a working group on petty trade as part of a sustainable planning process sought to strengthen trader associations, which, at that time, had excellent access to local government officials. By 1997 about 240 self-help groups represented 16,000 traders and two umbrella groups were established – VIBINDO (Association of Small Businesses) and KIWAKU (an association of clothes sellers).

VIBINDO started in 1995 with the aim of finding suitable plots of land for traders through negotiation with local authorities, but their initial attempts failed as the land was usually squatted before formalities could be finalized; the organisation then focused on advocacy and training. VIBINDO achieved considerable status and, by 2007, represented about 300 associations with a combined membership of 40,000 people (Msoka, 2007).

In March 2006, a letter from the Prime Minister instructed urban local authorities throughout Tanzania to move street traders from busy areas, initiating widespread clearances. VIBINDO negotiated a stay of execution for six months, and tried to negotiate alternative sites, but these were not accepted and the evictions resumed on 30 September 2006 (key informant interviews). Police and militia were drafted in to demolish kiosks and temporary structures, to confiscate goods and prosecute traders. It is difficult to know how many traders were affected, but Lyons and Msoka (2010) suggest that up to a million traders may have lost their place of work through eviction or intimidation.

There is little political support to resolve the problems of street trade. For example, in 2008 the Mayor of Ilala took an initiative to close Lumumba Street on Saturday evenings for a night market; the government shortly closed it down. Evictions and harassment remain very significant. Of the 153 traders who answered the relevant questions, 80% reported suffering harassment and 71% evictions. Yet, traders in the survey considered they had a claim and rights to trade: 'I'm Tanzanian and not a refugee,' one claimed, while others argued they had rights because they paid daily fees. The evictions are punitive and have far-reaching consequences, as this trader reported:

> I'll never forget the day when the police attacked hawkers at Mwenge. They grabbed my goods ... I was also injured on my back... Till now I've got scars on the back. I asked myself, 'What does the government want us to do?' I urge the government create new places for the hawkers to trade because eviction does not help us to get out from this extreme poverty.
>
> Male trader, selling second-hand clothing,
> Mwenge bus stand

Despite the wholesale clearances of 2007 and subsequent clearances in 2011, for new infrastructure projects, traders have some political clout; their voting potential is important and they argue they have rights to survive. Thus, before elections, politicians quietly turn a blind eye to vending, and a *laissez faire* approach is adopted (key informant interviews).

Towards collective rights in the public domain

The ongoing persecution and exclusion of informal economy workers in public space, with street traders as a prime target, suggests an urgent need for definition a new 'rights regime', that sees the public domain of cities as part of the urban land resource to which collective property rights may pertain. The usual conception of this public domain is that the land pertains to the state, with its functions limited to the use for highways and traffic circulation, infrastructure or public space, providing 'public goods' over which no individual or communal rights may hold sway.

This conception is far from the reality in most urban areas, particularly in cities with a large informal economy workforce, where streets provide an important place of work for the urban poor. Here many different rights regimes are found. Despite contravening local bylaws, traders often pay daily fees to kinsmen, municipal authorities, market managers, trading associations, adjacent property owners or other gatekeepers. The payment sometimes, but not always, confers a tenuous acceptance of legitimacy. Often traders are most secure in the run-up to elections, when they are seen as voting banks for a dominant party, but this tenuous legitimacy can be overturned at whim by presidential decree or ministerial letter. Evictions, ranging from low-level displacements for a few hours to brutal repossessions, are commonplace.

The definition of urban public space as a *common-pool resource*, according to Ostrom's analysis, provides the promise of a new framework of collective rights in the public domain. These would include a variety of rights – to work, to move through and to enjoy public space, which should be open to all comers building on social structures and acceptable local practice. This has similarity to the *common heritage* approach that Delville suggests. The challenge is how to manage the resulting congestion and 'tragedy of the commons' that may occur. For street traders, the bundles of rights might include access and beneficial use, but with conditions requiring contributions to collective management of the resource. A complex range of overlapping rights could be envisaged, for example, to provide for time-sharing of space. Local management would be essential and, although information requirements would be complex to maintain, traders are increasingly learning new technologies.

As Dakar has shown, political support creates space for accommodation in which new management structures and approaches may occur. It has not yet, however, resulted in radically new solutions to space management – the common policy agenda of moving traders to poorly located purpose-built sites, as proposed in rue Félix Eboué in Dakar, is almost always a failure. This is vividly illustrated by the spectacle of abandoned corridors in the Machinga Complex (a hawkers' complex) in Dar es Salaam.

It is clear from the case studies that traders' associations play a vital role in negotiating space. In both cities the old-established associations, UNACOIS in Dakar and VIBINDO in Dar es Salaam, have been co-opted by the establishment and are too afraid of losing influence to protect traders in case of a violent eviction. More transparency in the process of negotiations would help. In Dakar FAMATS is now linked to an international network of street traders, bringing solidarity and new ideas. However, establishing property rights is only one aspect of establishing legitimacy for those in the informal economy. While their activity remains illegal across a number of spheres, they will remain vulnerable.

References

Alden W.L. (2002) Formalizing the informal: is there a way to safely unlock human potential through land entitlement? A review of changing land administration in Africa, in Guha-Khasnobis, B., Kunbur, R. and Ostrom, E. (eds), *Linking the Formal and Informal Economy: Concepts and Policies*, Oxford: Oxford University Press.

Benkahla, A. and Seck, M. (2011) Pour une véritable concertation sur les enjeux et objectifs d'une réforme foncière au Sénégal, Initiative Prosective Agricole et Rurale (IPAR), http://www.inter-reseaux.org/IMG/pdf/Policy_brief_-_reforme_fonciere-4.pdf, accessed September 2011.

Boye, A.-K. (1978) Le regime foncier sénégalais, *Ethiopiques* (14), http://ethiopiques.refer.sn/spip.php?article645, accessed September 2011.

Brinkerhoff, D. and Goldsmith, A. (2005) Institutional dualism and international development: a revisionist interpretation of good governance, *Administration and Society* 37(2): 199–224.

Brown, A. (ed.) (2006) *Contested Space: Street Trading, Public Space and Livelihoods in Developing Cities*, Rugby: ITDG.

Brown, A., Lyons, M. and Dankoco, I. (2010) Street traders and the emerging spaces for urban voice and citizenship in African Cities, *Urban Studies*, 47(3): 666–83.

Dankoco, I. (2011) Rapport d'étude documentaire et qualitative: régulation du commerce de rue à Dakar, unpublished research report.

Davis, M. (2006) *Planet of Slums*, London: Verso.

de Soto, H. (1989) The Other Path: The Invisible Revolution in the Third World, New York: Perennial.

de Soto, H. (2000) *The Mystery of Capital: Why Capitalism Triumphs in the West and Fails Everywhere Else*, New York: Basic.

Delville, L. (2000) Harmonising formal law and customary land rights in French-speaking West Africa, in Toulmin, C. and Quan, J. (eds), *Evolving Land Rights, Policy and Tenure in Africa*, London: DFID/IIED/NRI: 97–121.

Delville, P. (2006) *Registering and Administering Customary Land Rights: PFRs in West Africa*, http://siteresources.worldbank.org/RPDLPROGRAM/Resources/459596-1161903702549/S4_Lavigne.pdf, accessed September 2011.

Demsetz, H. (1967) Towards a theory of property rights, *American Economic Review*, 57(2): 347–59.

Durand-Lasserve, A. and Royston, L. (eds) (2002) *Holding their Ground: Secure Tenure for the Urban Poor in Developing Countries*, London: Earthscan.

Fernandes, E. and Varley, A. (1998) Law, the city and citizenship, in developing countries: an introduction, in Fernandes, E. and Varley, A., *Illegal Cities: Law and Urban Change in Developing Countries*, London: Zed.

Gilbert, A. (2002) On the mystery of capital and the myths of Hernando de Soto: *what difference does legal title make?*, *Town Planning Review*, 24(1): 1–19.

Hardin, G. (1968) The tragedy of the commons, *Science*, 162 (13 December): 1243–8.

Harvey, D. (2003) Debates and developments: the right to the city, *International Journal of Urban and Regional Research*, 27(4): 939–41.

Hesseling, G. (2009) Land reform in Senegal: l'Histoire se répète?, in Ubink, J., Hoekema, A. and Assies, W., *Legalising Land Rights: Local Practices, State Responses and Tenure Security in Africa, Asia and Latin America*, Leiden, Netherlands: Leiden University Press: 243–71.

Hillbom, E. (2011) The right to water: an inquiry into legal empowerment and property rights formation in Tanzania, in Banik, D. *The Legal Empowerment Agenda: Poverty, Labour and the Informal Economy in Africa*, Farnham and Burlington, VT: Ashgate: 193–213.

Joierman, S. (2001) Inherited legal systems and effective rule of law: Africa and the colonial legacy, *Journal of Modern African Studies*, 39(4): 571–96.

Kironde, L. (2000) Understanding land markets in African urban areas: the case of Dar es Salaam, Tanzania, *Habitat International*, 24(2): 151–65.

Klerman, D., Mahone, P., Spamann, H. and Weinstein, M. (2010) Legal origin or colonial history?, https://papers.ssrn.com/sol3/papers.cfm?abstract_id=1903994, accessed September 2011.

Kombe, W. (2000) Regularizing housing land development during the transition to market-led supply in Tanzania, *Habitat International*, 29(2): 167–84.

Low, S. and Smith, N. (2005) *The Politics of Public Space*, New York and Abingdon: Routledge.

Lyons, M. and Msoka, C. (2010) The World Bank and the street: (how) do 'Doing Business' reforms affect Tanzania's micro-traders?, *Urban Studies*, 47(5): 1079–97.

Lyons, M. and Snoxell, S. (2005) Sustainable urban livelihoods and marketplace social capital: crisis and strategy in petty trade, *Urban Studies*, 42(8): 1301–20.

Mallya, E. (2005) Women NGOs and the policy process in Tanzania: the case of the Land Act of 1999, *African Study Monographs*, 26(4): 183–200.

McAuslan, P. (1998) Urbanization, law and development: a record of research, in Fernandes, E. and Varley, A., *Illegal Cities: Law and Urban Change in Developing Countries*, London: Zed.

Merry, S.E. (1988) Legal pluralism, *Law and Society Review*, 22(5): 869–96.

Msoka, C. (2007) An assessment of the informal economy associations in Tanzania: the case study of VIBINDO society in Dar es Salaam, Paper to the conference on Informalising Economies and New Organising Strategies in Africa, Nordic Africa Institute 20–22 April.

Odgaard, R. (2003) Scrambling for land in Tanzania: process of formalisation and legitimisation of land rights in Tanzania, in Benjaminsen, T. and Lund, I., *Securing Land Rights in Africa*, London: Frank Cass: 71–88.

Olenasha, W. (2004) Reforming land tenure in Tanzania: for whose benefit?, https://www.researchgate.net/publication/255570290_Reforming_Land_Tenure_In_Tanzania_For_Whose_Benefit, accessed January 2009.

Ostrom, E. and Hess, C. (2010) Private and common property rights, in Boukaert, B., Property Law and Economics, Cheltenham: Edward Elgar: 53–107.

Santos, B. de Souza (2006) The heterogeneous state and legal pluralism in Mozambique, *Law and Society Review*, 40(1): 39–75.

Shivji, I. (1998) *Not yet Democracy: Reforming Land Tenure in Tanzania*, London: IIED; Dar es Salaam: HAKIARDHI Faculty of Law, University of Dar es Salaam.

Thioub, I., Diop, M.-C. and Boone, C. (1998) Economic liberalisation in Senegal: shifting politics of indigenous business interests, *African Studies Review*, 41(2): 63–89.

UNDP (2008) *Making the Law Work for Everyone: Volume 2, Report of the Commission on Legal Empowerment of the Poor*, New York: UNDP.

URT (1997) National Land Policy, URT (United Republic of Tanzania, Ministry of Lands and Human Settlements Development), http://www.tzonline.org/pdf/nationallandpolicy.pdf, accessed April 2016.

URT (1999a) The Land Act, 1999 (Act No. 4/1999), URT (United Republic of Tanzania), http://www.ecolex.org/details/legislation/land-act-1999-no-4-of-1999-lex-faoc023795/, accessed September 2011.

URT (1999b) The Village Land Act, 1999 (Act No. 5/1999), URT (United Republic of Tanzania), http://faolex.fao.org/docs/pdf/tan53306.pdf, accessed September 2011.

URT (2004) The Land (Amendment) Act, 2004 (Act No. 2/2004), URT (United Republic of Tanzania), http://faolex.fao.org/docs/pdf/tan53051.pdf, accessed September 2011.

URT (2008a) The property and business formalisation programme, reform designs, Volume 1: The Executive Summary, Cost-benefit analysis, http://www.tanzania.go.tz/mkurabita1/mkurabita_report_index.html, accessed September 2011 [site no longer live].

URT (2008b) The property and business formalisation programme, reform designs, Volume 2: Property formalization reforms outlines and packages for mainland Tanzania, Cost-benefit analysis, http://www.tanzania.go.tz/mkurabita1/mkurabita_report_index.html, accessed September 2011 [site no longer live].

Webster, C. (2007) Property rights, public space and urban design, *Town Planning Review*, 78(1): 81–102

Webster, C. and Lai, W.-C. (2003) *Property Rights, Planning and Markets: Managing Spontaneous Cities*, Cheltenham: Edward Elgar.

World Bank (2016a) Senegal data, http://data.worldbank.org/country/senegal, accessed September 2011.

World Bank (2016b) Tanzania data, http://data.worldbank.org/country/tanzania, accessed September 2011.

7 Law and the informal economy

The WIEGO law project

Christine Bonner

Summary

For many informal workers – especially the self-employed – a key focus of their struggle for rights, protection, decent work and livelihoods is defending themselves against legal exclusion; challenging and changing existing laws; and ensuring the fair and effective implementation of laws and regulations. Yet often they do not have the information, capability, resources, access, organisational strength and negotiation fora to engage effectively in these processes. This chapter introduces the WIEGO Law and Informality project, designed to strengthen workers' voice in tackling this challenge.

Introduction

In 2007, the WIEGO[1] network (WIEGO, 2016) launched a project on Law and Informality as a pioneering way to meet the enormous demand from informal workers, to better understand, use and reform the legal system in order to establish more inclusive economies and more resilient societies. The Law and Informality project documented how laws and regulations – or their absence – affect different groups of informal workers. The project identified labour legislation, environmental laws, municipal laws, sector-specific laws, right-to-information laws and property laws as having significant impact on different sectors in the informal economy.

The project began as a pilot in India and then expanded, in 2011, first to Colombia, then to Ghana, Peru and Thailand. In 2013 a new phase included South Africa and, once again, India. The project was designed to understand which laws and regulations impinge on the working poor in the informal economy, either negatively or positively; the nature of the legal demands, struggles and strategies of organisations of informal workers; which of these have been successful; and where are there good examples of appropriate legislation and litigation.

The overall objective of the project was to make a significant contribution to the development of an enabling legal and regulatory environment for informal workers, especially women, which promotes laws, regulations and policies that provide protection and support to these workers and helps build the capacity of organisations of informal workers to engage effectively in processes leading to legal reforms and their legal empowerment. The specific objectives were:

- to document and analyse laws/regulations/policies that impinge on the work, working conditions and livelihoods of four different occupational groups of informal workers;
- to identify useful laws, regulations and judgments (better practices), and key legal strategies and struggles of these informal workers that can be widely shared;
- to create a platform of demands and model clauses, laws and agreements that are useful for informal worker organisations in their engagement with authorities or employers;
- to help build the capacity of informal worker organisations, and especially women leaders, to engage with legal issues, through advocacy, negotiation and implementation of favourable legal change;
- to contribute to conceptual change concerning notions of 'work', 'worker' and 'work organisation'.

Phase 1, funded by the International Development Research Centre (IDRC), had a number of major achievements. The project set up a Law and Informality site (WIEGO, 2016), which contains an observatory of laws, judgments, legal opinions, videos and reports on projects in different countries, exposure dialogues, legal briefs and accessible materials for workers, lawyers, researchers and policymakers. The project identified and reported on general laws and regulations applicable to informal workers, and those applicable to different sectors in three countries. It conducted legal capacity-building activities and supported informal worker organisations in their advocacy and campaigning around legal change and implementation.

Phase 2 of the project from 2013–14, funded by the Ministry of Foreign Affairs of the Netherlands under the Funding Leadership Opportunities for Women (FLOW) programme, sought to support and build the capacity of 12 member-based organisations (MBOs) in five countries, across four sectors, to access information about the laws and regulations that affect them, develop plans for advancing their rights and the economic viability of their work, and make the legal framework more supportive of their livelihoods. By September 2014, a total of 2,843 informal economy workers had participated in activities promoted by the programme. A list of legal allies was compiled from the five countries. Reports from the five countries include a list of at least 98 laws and legal instruments applicable to the groups of informal economy workers on the project. These deal with street vendors, home-based workers, domestic workers and waste pickers.

The generic outputs of the project include: information on the legal situation of informal workers; active ongoing engagement on legal reforms between organisations of informal workers, legal experts and relevant government officials; and strengthened capacity of organisations of informal workers to understand the laws and regulations affecting them and to engage with relevant authorities.

During the course of the project, several changes to laws and regulations took place, including the adoption of a Homeworkers Protection Act (2010) and a Domestic Workers Ministerial Decree (2012) in Thailand; a draft Law of the Self-Employed Worker (under active review by the Ministry of Labour) in Peru; a new Metropolitan Ordinance regulating street vending in Lima, Peru; and a Task Force on Domestic Workers and Accra Municipal Authority review of possible laws or regulations for domestic workers and street vendors, respectively, in Ghana.

By 2014, 164 informal workers from four occupational sectors from Asia, Africa and Latin America had demanded improvement of labour rights in the context of International Labour Office (ILO) discussions on 'the transition from the informal to the formal economy', through regional and local workshops. At the International Labour Conference (ILC) in 2014, 15 women workers engaged directly in the ILO negotiations and three informal worker leaders spoke at ILO Assembly Hall. Half of the amendments and key issues put forward by the WIEGO network group and allies were adopted. In 2015 the delegation again participated in the ILC for the final session of the transition negotiations, which resulted in Recommendation 204 Concerning the Transition from the Informal to the Formal Economy, the first ILO standard to cover all informal workers.

In October 2015, recognising the importance of law for informal workers and its applicability across all WIEGO programmes, WIEGO launched a new programme on Law and Informality, as law is of such importance to informal workers. The programme aims to 1) analyse and improve legal and regulatory frameworks for informal workers and 2) empower workers' organisations to fight for their recognition and rights as workers. The programme builds on the work and achievements of the first Law and Informality project.

Context

The informal economy is widely seen as the set of economic units, activities and workers that are outside the reach of government regulation; and informal operators and workers are widely understood to be trying to evade or avoid regulations and taxes. However, these notions need to be re-examined for three important reasons. First, many informal workers do not choose to avoid or evade the law: they are simply trying to earn an honest living. When there is no appropriate legal framework, they do not know the legal procedures, or find the procedures too cumbersome to comply with, informal workers are often forced to operate outside the law. Second, those who work in unregulated enterprises or unprotected jobs are often *not* outside the reach of the regulatory arm of government. Because they are perceived to be illegal, informal workers are often subjected to criminal law or punitive actions by government and police. Third, informal workers are outside the reach of government protection, not just regulation.

Consider the situation of street vendors in India. Municipal governments have the duty and task of managing public spaces in such a way that they can be used for different purposes: notably, street vending as well as pedestrian and

Figure 7.1 Most cities do not fairly allocate or adequately regulate public space – to everyone's detriment.

vehicular traffic. Municipal Corporation Acts set out the duties, responsibilities and power of municipal governments, including improving streets and keeping them free from obstruction, and issuing licences for selling in public spaces or markets. Urban planning and development acts deal with the allocation of land for development. But most cities have not issued new licences or permits for many years and do not fairly allocate or adequately regulate public spaces – to everyone's detriment, including the street vendors.

In this vacuum, street vendors can be charged on a number of counts – notably, for causing an obstruction – or have their goods confiscated or be evicted. The police in India can justify arresting, evicting, or otherwise penalising street vendors, especially those who do not have a licence or permit, on the basis of several national laws – for example, the Indian Penal Code, 1860, prohibits any act that creates a danger within a public right of way; the Criminal Procedure Code, 1973, allows a police officer to arrest anyone about to commit any cognisable offence without orders from a magistrate and without a warrant, and the Motor Vehicle Act, 1988, seeks to ensure the free flow of traffic. The use of these laws to harass vendors has devastating consequences for their livelihoods. Extensive lobbying by the National Association of Street Vendors of India (NASVI) has resulted in the enactment of ground-breaking Federal legislation, the Street Vendors (Protection of Livelihood and Regulation of Street Vending) Act, 2014, which provides an enabling environment for street vending (see Chapter 8). The law project was involved in capacity-building to help vendors understand the new law.

A key lesson from the work is that the options for improving the lives and livelihoods of informal workers through legal channels are multifaceted:

• Sub-contracted and own-account workers fall within a broad continuum of working arrangements. International instruments have recognised these categories as 'workers' even though they fall outside of a traditional employer–employee relationship.

- A broader definition of 'worker' must recognise the economic dependency of those outside a traditional employer–employee relationship, enabling dependent workers who are not employees to obtain claims and benefits from the commercial commodity chain.
- Imposing obligations (such as levying a tax) on traders and consumers would recognise the economic dependence of those workers who are neither employees nor independent contractors. Human rights principles and the need to ensure decent work could also be used to impose a welfare obligation on the public.
- Workers who do collecting (such as fish and forest workers) could receive prices for collected goods that reflect the value of the labour and the knowledge used in collecting.
- Including those involved in primary collection (e.g. waste, natural resources) and not just those in the final stages of production or value addition as workers within an industry would have far-reaching benefits across all sectors.
- The right to access public resources (recognised in Recommendation 204) – whether urban spaces, waste or natural resources – is fundamental across all sectors; key legal battlegrounds pit property rights and privatisation against the demands of informal workers for the right to livelihood and social security.

A common demand that all groups of informal workers have pursued is social protection. Legislators and policy-makers have generally expressed more receptiveness to this demand than to the other core demands for employment rights. However, the most challenging and contentious of the informal workers' legal demands are those pertaining to working conditions, including secure tenure. Finally, a key lesson from the work is that different normative or legal frameworks and different branches of law should be and have been invoked to protect the rights of informal workers.

Securing livelihoods of the working poor in the informal economy requires focused and sustained attention on all policies and legislation that have direct or indirect impact on the workers in different occupational groups. Thus, while broad demands for strengthening labour rights and working conditions are maintained, sector-specific demands of each occupational group need to be pursued as well. This requires parallel strategies of *using* the law and *changing* the law. For example, at the same time that workers' groups are using the international human rights norms enshrined in ILO Conventions and national constitutions, they are also working to change those norms to better reflect the realities of the informal economy through strategies such as working for General Comments from the Human Rights Committee or ILO rulings that explicitly mention informal workers, in order to be able to cite an emerging international norm in subsequent advocacy work. Similarly, when WIEGO and its partners engage with a specific branch of law, like labour law, they try to encourage consistent and fair enforcement for informal workers where reasonable laws exist, while simultaneously working for better laws that contribute to an enabling regulatory framework through advocacy efforts, litigation and legislative reform.

In brief, in engaging with the law in support of the working poor in the informal economy, WIEGO and its partners acknowledge that law can be both a tool to take into other battles *and* a site of contestation: this perspective and approach is a key value added of the integrated Law and Informality programme.

Normative frameworks

In the past, the following normative frameworks have been invoked to protect the rights of informal workers.

Universal Declaration of Human Rights: Under Article 23 of the Universal Declaration of Human Rights, everyone has the right to work, to free choice of employment, to just and favourable conditions of work and to protection against unemployment; to equal pay for equal work; to just and favourable remuneration ensuring an existence worthy of human dignity, and supplemented, if necessary, by other means of social protection; and the right to form and to join trade unions for the protection of their interests.

International and regional human rights treaties: Major international human rights treaties protect a number of rights relevant to informal workers. For example, the International Covenant on Economic, Social, and Cultural Rights (ICESCR), 1966, protects the right to work (Article 6) and to just and favourable conditions of work (Article 7) and the International Covenant on Civil and Political Rights (ICCPR), 1966, protects against arbitrary search and seizure (Article 9) and equality before the law (Article 26). The Convention on the Elimination of All forms of Discrimination Against Women (CEDAW), 1979, and the International Convention on the Elimination of All Forms of Racial Discrimination (ICERD), 1965, also contain provisions relevant to informal workers in many countries. Additionally, regional human rights treaties, such as the African Charter on Human and Peoples' Rights, 1986, the European Convention on Human Rights, 1953, and the American Convention on Human Rights (San José Pact), 1978, have also been used to argue in favour of the rights of informal workers.

International labour standards and conventions: Many international labour standards and conventions are designed for formal wage workers in a recognised employer–employee relationship, not for the self-employed or informal wage workers. However, international labour standards that embody fundamental human rights are to be enjoyed by all workers: including, the right to freedom from discrimination and the right to organise. Moreover, a number of ILO instruments apply explicitly to 'workers' rather than the legally narrower term 'employees', or do not contain language limiting their application to formal wage workers.

National constitutions: In many countries around the world, the constitution establishes rights to life, personal liberty and equality. Some also include the rights to survival or to pursue a trade or business.

National legislation: A growing number of countries have developed laws or legislation for specific groups of informal workers – notably, domestic workers, home-based producers and street vendors – or for all workers in specific industries.

Administrative procedures: In several countries, judges and lawyers have argued for the rights of informal workers on the basis of transparent administration procedures or principles of legitimate trust.

Branches of the law

In their legal negotiations, some groups of informal workers – notably, domestic workers and home-based workers – have focused on obtaining recognition as workers and pushing for regulation of working conditions, as well as social protection. However, other groups of informal workers have focused on other branches of law or other regulatory frameworks that have a direct impact on their livelihoods. For instance, street vendors focus on access rights to public space and security from evictions. Resource-dependent workers such as forest and fish workers focus on legislation that determines their access to resources, such as forest laws or coastal regulations. Similarly, waste pickers focus on municipal regulations, schemes and contracts that determine their access to recyclable waste materials. Several branches of legislation are particularly important for informal workers.

Urban planning and development: Urban planning and development acts deal with the allocation of land for development purposes. Urban public space – streets, sidewalks, parks and other open spaces – should be considered and treated as 'urban commons' to be accessed and used by the public – for livelihoods, as well as other purposes. It is important to note that urban commons are not 'true commons' in that they are usually state-owned, but the public has a right of access. For this reason, the management of urban commons cannot be left to cooperative management by the public: it requires the state – usually municipal governments – to regulate and manage the commons, not to privatise them.

There are two major legal challenges in regard to managing urban commons – that is, streets, sidewalks, parks and open spaces. The first is to persuade municipal governments that public spaces are common spaces. Although the public does not own them, they should not be privatised, as is happening widely in many cities, and the public should be allowed open access. The second legal challenge is to persuade municipal governments to manage and regulate the urban commons. In India and elsewhere, the courts have a right to intervene if the state – that is, municipal governments or the mayor's office – impose unreasonable restrictions on the use of public space. This responsibility should not be delegated solely to the police to maintain law and order, but municipal governments should introduce regulations and practices which take into account and balance the competing rights of different public groups – for example, street vendors and hawkers, pedestrians and vehicle owners/drivers.

Social protection legislation: In some countries, targeted social protection schemes have been introduced to protect informal workers. In India, the 2008 Unorganised Workers Social Security Act has the potential to cover all 'workers', including the self-employed (both dependent and independent), for the purposes of ensuring access to basic social security, while working hours, safety and employment relations continue to be unregulated for these workers. In other countries, universal schemes, designed to cover both formal and informal workers, have been introduced.

Sector-specific regulatory laws: A complex range of sector-specific regulatory laws impact workers in the informal economy, especially own-account workers and the self-employed, including laws relating to the use of public resources such as urban space or forests. A sectoral approach is necessary to enable critical understanding of the legal demands raised in each sector. The sector-specific laws or regulations that impinge on workers in the informal economy can be broadly categorised as follows. Regulations that:

- balance conflicting needs and users – for example, the rights of street vendors to use pavements trade and the rights of citizens to access pavements;
- determine access to, and sustainable use of, resources – for example, access of waste pickers to waste; and encourage composting and recovery of recyclables, rather than incinerators;
- determine access to government tenders and support services – for example, enabling associations of waste pickers to bid, alongside private companies, for solid waste management contracts; determining which enterprises get access to subsidies, tax holidays and trade promotion;
- impact markets and pricing – for example, governing access to credit, markets and support prices.

Labour law: The contract of employment has been the primary means through which a person is recognised as an employee and is granted the benefits and protection of employment. One challenge has been to extend the definition of 'employee' to those who are not directly employed by an employer or enterprise (e.g. contract and sub-contracted workers) and to those who appear self-employed but who display characteristics of economic dependency akin to employees. People who share 'employee-like' characteristics seek to be brought within the province of labour law in order to secure social protection and to have their working conditions regulated. Domestic workers have been most successful in getting their status as workers recognised under specific labour laws or welfare legislation.

Key areas of law affecting street vendors

Three areas of legislation specifically affect street vending. First, many cities have regulations or bylaws on street vending and hawking, but these often focus on control or restriction of these activities, rather than on enabling livelihoods and establishing measures to address conflicting demands – for example, over space. Second, licensing regulations, either ones that are generic or specific to vendors and hawkers, often affect street vendors. Licensing is usually undertaken by municipal governments, but there are many problems in implementation – for example, the number of permits issued may be far below the demand/need; the application form may be complex and require information that vendors cannot easily obtain (e.g. a postal address) and the cost of a licence may be set too high for street vendors to attain. Third are regulations relating to public rights of way and public order which are frequently invoked to restrict street vending. Critical problems for street vendors include frequent harassment or confiscation of goods and evictions, but in several cities organisations of workers have been able to lobby for effective changes to the law (see Chapter 8 on Ahmedabad) or challenge the restrictive regulations through court proceedings (see Chapter 9 on Durban).

Note

1   WIEGO (Women in Informal Employment: Globalizing and Organizing) is a global network focused on securing livelihoods for the working poor, especially women, in the informal economy. WIEGO and its affiliates represent around 2 million workers worldwide.

Reference

WIEGO (2016) Law and Informality, http://wiego.org/law, accessed April 2016.

Part II

Street trading at the front line

8 Claiming urban space through judicial activism

Street vendors of Ahmedabad

Darshini Mahadevia and Suchita Vyas

Summary

In December 2009 groups and individuals affected by displacements came together for a Public Hearing on displacements in Ahmedabad City. Women street vendors from the Self-Employed Women's Association (SEWA) said, 'We have lost our identity: we need only 2 square feet to sell bread.' This chapter charts the events leading to the hearing, the conflicting and hostile legal context in which vendors operate, and how vendors used public interest litigation to stake their claim – with mixed outcomes.

Context

On 19 December 2009, people affected by displacements and concerned individuals in Ahmedabad participated in the first ever Public Hearing on Displacements of Lives and Livelihoods in Ahmedabad City (OIA, 2010). The context was the political vision of World City status for Ahmedabad that has shaped its recent development, promoting 'vibrant' economic development, a new Bus Rapid Transit system (BRT), mega-malls, urban core and riverfront renewal, lake development and suburban gated communities. For the urban poor, the outcome of these transformations has resulted in displacement of lives and livelihoods. Slums were demolished and many residents relocated on the city's periphery in uninhabitable sites, while street vendors were displaced for flyover, upgrading and road-widening projects.

One of the most vociferous groups at the Public Hearing were speakers from the SEWA, the trade union of self-employed women representing nearly 2 million workers, including vendors, campaigning for the rights of street vendors. Although street vendors already faced constant harassment from municipal government and the police, SEWA argued that the development projects had created new hardships for vendors as trading space was lost to infrastructure and could not be reclaimed. Ahmedabad Municipal Corporation (AMC) had offered no alternate, and the women vendors argued that their 'natural markets' were being destroyed. As Revaben Narsinghbhai Vaghela said, 'What is our identity? We need only a space for two *topali* (large bamboo baskets)' (ibid.).

The concept of 'natural market' had been conceptualised by SEWA, and was enshrined in the National Policy on Urban Street Vendors, 2004 (NPUSV), the federal policy adopted by the Ministry of Housing and Urban Poverty Alleviation (MHUPA) after years of campaigning by the National Association of Street Vendors of India (NASVI). The NPUSV defined a natural market as a place where a confluence of pedestrian and vehicle movements provides an excellent location to trade. The policy was updated in 2009, but was not widely adopted, and NASVI successfully lobbied for the introduction of legislation, now enacted as the Street Vendors (Protection of Livelihood and Regulation of Street Vending) Act, 2014 (GoI, 2014), federal legislation that must be implemented by the Indian states.

As Revaben indicated, natural markets by schools or bus-stands are places where a demand for goods emerges and vendors congregate, but when a market is dispersed vendors scatter, losing both clients and income. As the Public Hearing report states, in 2009 construction of a flyover at Hatkeshwar in east Ahmedabad dispersed about 350 vendors, reducing their incomes from about Rs 70–150/day to Rs 20–40/day (OIA, 2010: 17). As the vendors argued, 'We are not against development, so why are our livelihoods being attacked?' Some had also lost homes to road-widening projects. Nationally, Ahmedabad has been highly successful at winning funding under the Jawaharlal Nehru National Urban Renewal Mission (JNNURM). Phase 1 of the JNNURM was a seven-year, US$20 billion urban improvement programme introduced by the federal government, which ran from 2005–6 to 2012–13, and won several awards for its work. This chapter, however, argues that renewal and redevelopment in Ahmedabad has been largely detrimental to the poor.

The research for this chapter was undertaken in 2011–14, funded under the Economic and Social Research Council Department for International Development (ESRC/DFID) project on law, rights and regulation for the urban informal economy, based on a detailed legislative review, a survey of 200 vendors and interviews with community and human rights organisations, AMC and many others. Key findings are discussed below. The research

Figure 8.1 Natural markets are where a demand for goods exists: Khodiyarnagar and SEWA market leaders Ahmedabad, India.

highlighted the volatile and starkly contrasting development scenarios in Ahmedabad; on the one hand, lives and livelihoods of a section of population were being damaged, while, on the other hand, the city was receiving laurels for 'development'.

This chapter examines the context of Ahmedabad, and the city's political economy and effect on street vendors, including the impact of displacements on their businesses and incomes, as highlighted in the survey. The chapter then examines the prohibitive legal context of street vending in Ahmedabad, and details the measures taken by SEWA which fought a three-year public interest litigation (PIL) battle against AMC in the Gujarat High Court, with mixed results; it then presents the conclusions and recommendations from this study.

Ahmedabad: contested visions of development

Ahmedabad is the largest city in Gujarat, one of the fastest-growing states in India, with GDP growth averaging of 8.4% *per annum* from 2004–5 to 2009–10 (Mahadevia, 2013). Although Gujarat is considered a model state by some researchers for its neoliberal mode of economic development (Debroy and Bandari, 2011), others highlight the lopsided development, and the priority for economic growth over human development, equity and environmental sustainability, which is embedded in the state's development paradigm (Hirway and Mahadevia, 2004).

Greater Ahmedabad had a population of 6.35 million in 2011, with an annual population growth of 3.5% (MoHA, 2011). However, in recent years, the city has changed its development trajectory. Although once the site of Mahatma Gandhi's Ashram and setting for the Textile Labour Association of Ahmedabad, it has since seen religious and ethnic conflict, witnessing communal riots during the 1980s and early 1990s and violence against the Muslim community in 2002, which resulted in the city's segmentation by religion (CCT, 2002; Mahadevia, 2007). The city is also divided by class, with industrial land in the east and high-class residential development in the west.

Mahadevia (2010) has argued that Ahmedabad is now a regressive city, which moved from the Gandhian ideology of benefit to all and successful participatory programmes such as the Slum Networking Programme (SNP) to an exclusive and non-participatory city. For example, the city has not implemented the decentralised governance envisaged under the 74th Constitutional Amendment (CA), 1992, or JNNURM governance reforms (Mahadevia, 2010), and has subverted the intent of the NPUSV, as described in this chapter. The state's rapid economic growth through capital-intensive industrialisation is a political agenda in which the state government has superseded AMC in policy-making, undermining the autonomy of the city government (ibid.). The outcome has been a state oblivious of the travails of the urban poor living and working in the informal sector. For example, since 2005 construction for the new BRT has led to a spate of demolitions and evictions.

Since the decline of Ahmedabad's textile industry in the late 1980s, informal sector employment has increased rapidly. Self-employment among men increased from 34.7% in 1987–8 to 53.6% in 2009–10, and among women from 38.2% in 1987–8 to 49.2% in 2009–10. In parallel, salaried employment for men decreased from 44.9% in 1987–8 to 37.8% in 2009–10, although for women it has remained relatively constant at around 30% (Mahadevia, 2012: 20). The shift has been to the services sector, which tends to be low paid, and to outsourcing and informalisation of manufacturing, and 48% of men and 91% of women working in manufacturing were self-employed in 2009–10 (ibid.: 22).

Although there has been no census, numbers of street vendors are reported to have increased, as now recognised by the 2009 NPUSV, which stated that 'street vendors form a very important segment of the unauthorized sector in the country', estimating that they account for about 2% of the working population in some cities, and employ large numbers of women (MHUPA, 2009). Their role in providing both 'affordable' and 'convenient' services to urban populations was also recognised.

Subversive claim: the street vendors of Ahmedabad

Street vending is found throughout Ahmedabad; a 2003 study for SEWA identified 164 'natural markets' in the city (Dalwadi, 2003). Figure 8.2 shows the location of these markets, also identifying those likely to be dispersed for development projects.

Our research was conducted in ten markets with different trading characteristics, three within and near the historic walled city (Bhadra, Delhi Darwaja and Jamalpur), one in south Ahmedabad around a improved lake frontage (Kankaria Lake), one in east Ahmedabad along the BRT route (Khodiyarnagar) and five in west Ahmedabad (Parasnagar, Nehrunagar, Vastrapur, near the Indian Institute of Management (IIM) and near CEPT University).

Bhadra market in the walled city was the oldest market, a cluster of about 5,000 vendors selling T-shirts, clothing and household goods. The market was to have been displaced by the JNNURM-funded Walled City Revitalization project. There was no policy statement from AMC about reallocation of vending space. On Sundays, the market overflowed to the riverbank for the weekly Gujri market, although this was removed for the Sabarmati Riverfront Development Project. By 2016, several hundred vendors had recolonised the newly paved area between Bhadra Fort and the historic gate, Teen Darwaja, and negotiations with AMC were ongoing.

In Delhi Darwaja vendors mainly sold used clothes. Jamalpur market was linked to the wholesale vegetable and fruit market nearby. At Kankaria Lake, vendors were dispersed for the gated lake-front tourist development, but continue to sell toys, sweets and food just outside the gates – many of these vendors were also evicted from surrounding slums and now live in squatter settlements on the urban periphery. Khodiyarnagar market mainly sold fruit and vegetables with some cooked food. In west Ahmedabad, at Parasnagar, a peripheral market, vendors sold vegetables and fruit; IIM and Vastrapur sell cooked food, and Nehrunagar and CEPT clothes and accessories.

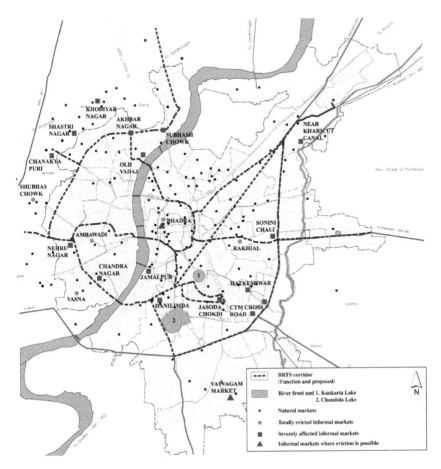

Figure 8.2 Ahmedabad: map of natural markets and potential displacements.
Source: OIA, 2010.

Of the 200 street vendor interviews, about 20–40 vendors were selected from the six large markets (Bhadra, Delhi Darwaja, Jamalpur, Kankaria Lake and Parasnagar), with smaller samples in the other four markets to test pilot hypotheses.

Gender and ethnicity

Of the 200 vendors interviewed, 70 (35%) were women vendors, who dominate in specific trades – for example, in Delhi Darwaja in the collection and sale of used clothes (17/30 market interviews) and at Jamalpur in the sale of vegetables and fruits (15/19 market interviews). Bhadra market is overwhelmingly male-dominated, and 60% of vendors interviewed were Muslims. Two-thirds of the whole sample were from scheduled castes (SCs) and other backward castes (OBCs), mainly from the Devipujak and Patni communities, who are traditionally vendors.

Age and education

About half the vendors interviewed were aged 35–54 years, 25% were aged 25–34 years and, for most, vending was their main household income. In Ahmedabad, vending is a long-term activity and about 43% of vendors interviewed had been vending for 20 years and more, with only 20% starting the business in last five years. Some 28% were illiterate and another 16% could only sign their names. In the two markets dominated by women, there were higher levels of illiteracy – 63% at Jamalpur and 43% in Delhi Darwaja.

Poverty indicators

The economic condition of vendors' households is weak. Those surveyed had average household sizes of 6.4, as compared to 5.0 in Ahmedabad as a whole, with an economic activity rate of 33%, compared to 36% for the whole city (Mahadevia, 2012: 19). Average net income from vending at the time of survey (2011) was reported as Rs 263/day or around Rs 7,000/month. Thus, for one worker, earning around the mean and supporting a household of six people, monthly per capita income would be Rs 1,166 – higher than the 2009–10 poverty line of Rs 951 per capita per month (India, Planning Commission, 2012). For most vendors interviewed, vending was their main job, with 80% working seven days a week, and 97% working for 12 months a year at the trade.

Access to space

Access to space is important for vendors, as 46% of respondents were stationary. The other 54% were mobile vendors, of whom most (83%) used a *lari* (a flat trolley usually 1.8m x 1.1m on four bicycle wheels). Of the stationary vendors, 72.5% displayed their wares on the ground or wall, the others used a wooden or metal stand or kiosk. Nearly three-quarters of those surveyed (73.5%) sold in the open, without an umbrella or shelter. Space-sharing was important; some markets open in the morning, such as Delhi Darwaja and Jamalpur, while the suburban markets start around 4pm. Only Bhadra was an all-day market. Lack of space therefore restricts vending hours. Nearly 80% of vendors had an informal arrangement to secure space, which often involved weekly or monthly payments, although new vendors found it extremely difficult to become established, particularly in the older markets of Bhadra, Delhi Darwaja and Jamalpur. One vendor in Bhadra reported paying Rs 50,000 as a returnable deposit and monthly rents of Rs 3,000 or Rs 3,500 during festivals to middlemen. He shared storage with four other vendors, which cost him Rs 750/- a month. Vendors also have to make to 'protection' or bribe payments.

Table 8.1 Bhadra, Ahmedabad: potential contribution of vendors to the local economy

Item	Number	Rs
Approx no. of street vendor enterprises in Bhadra (1)	3,500	
Average workers per business (survey findings)	2.5	
Estimated number of vendors in Bhadra	8750	
Average business turnover per day (Rs) (survey findings)		**1,880**
Average days per week worked (2)	6.97	
Weeks per year worked (2)	51.65	
Average working days per year	**360**	
Annual average turnover (3)		Rs 236 crore (Rs 2,368m)
Annual average turnover (US$) (4)		US$ 43.75 million

Notes: (1) Jajoo, 2012; (2) Data analysis; (3) 1 core = 10m, Rs 236 crore = Rs 2.36 bn; (4) Exchange rate US$ 1 = Rs 54.15 (this varies over time); (5) Figures in bold used to calculate annual average turnover for the market.

Economic value of street vending

The overall value of street vending to the urban economy is significant, although difficult to estimate because of uncertainties on income data and numbers of street vendors. However, the research team estimated that in Bhadra alone, which had about 3,500 vending businesses at the time of the survey, the economic contribution of the vendors was Rs 236 crore or US$ 43 million (Table 8.1). Vendors also supported other micro-businesses, including porters, guards, food and tea sellers (*chaiwalas*), repairs and other services, a fact rarely understood by policy-makers. Many vendors, particularly in Bhadra, paid an *aagyevan* (leader/ landlord) or other middleman, who may collect money for the bribes demanded by state actors to overlook this 'illegal activity'.

Vulnerabilities

Some interesting dynamics emerge from an analysis of vulnerability assessed against the reported aspects of harassment, fines, confiscation or eviction. Women vendors appear to be more vulnerable to confiscation and to suffer much higher rates of eviction than men (67% of women vendors reported an eviction compared to 40% of men), but men report a slightly higher rate of fines. Those with the least permanent facilities – trading on the ground or carrying their goods – also report significantly higher levels of eviction (60%) compared to those with a more established vending set-up (i.e. trading from a table, bed, *lari* or bicycle).

Muslim vendors were less subject to fines and harassment than Hindus and other religions, probably because most were trading in Bhadra, which has a predominance of Muslim vendors and the established *aagyevan* system. But, this area also reported high levels of eviction and confiscation, probably because most were trading in the historic city. Many vendors suffered extreme and ongoing problems. The stories of vendors demonstrate an entrenched history of harassment and eviction over a long period of time, often by police and city officials.

> I saw the police taking bribes from each of the carts with vegetables coming from the APMC market. They were collecting around Rs 20–30 from each cart. There was a police van standing outside gate with three to four police and each of the police took turns to stand at the gate to collect money.
>
> Female, Jamalpur, selling chilli and garlic

> The police and corporation people carried out *laathi*-charge (beating with a stick) and would sneak away goods if I can't pay *hafta* (bribes). Once a man objected to the *laathi*-charge, but the matter was taken to Ela Ben in SEWA, she fought against the police.
>
> Female, Manek Chowk, selling garlic,
> chilli and vegetables

Certain events trigger a crackdown – for example, the visits of high-level politicians. During festivals both rents and bribes escalate. Evictions were still ongoing in all the markets visited.

> Whenever any politician goes by we are not allowed to sit here and the police visit to evict us, even if there are no traffic problems. Once the police go back we come back.
>
> Male, Delhi Darwaja, selling
> second-hand clothes

> I created five to six stalls during the kite season to sell kites and threads.[1] At that time all the policemen from the Ghatlodia police station (from high to low rank) came and took goods worth Rs 5,000 from each stall as well as cash bribes. This is very unfair because we made our stall on somebody's land and we pay Rs 1,000 per stall. When the policemen take too much bribes then all our profit is eaten up.
>
> Male, Parasnagar, vegetable/kite
> seller near Utarayan festival ground

The stress of evictions is acute, and the effect of combined evictions from housing and livelihoods can be devastating.

> I stayed at Macchi Pir slum. I had gone to my village as my in-laws expired. I was given notice that this land is municipal land. When I came back my house was destroyed by a bulldozer. I lost all my documents except my voting card. Now I have no place to stay, so I stay on the road. My family has been vending here since generations. It was good there. They took photos and told us we would be given space to sit. After it was built we were allowed to

sit next to the boundary. Then a few months ago during Christmas they built a small gate and we were pushed out. Earlier we were 100 in this row. Now only 30 remain, the rest have gone to do labourers' work.

Male, Kankaria Lake, selling popcorn

It is clear that women vendors and those from SCs are particularly vulnerable, but that the economic contribution of vendors is considerable. Vending activity requires space, and vendors pay the middlemen and members of the local state – the police and local government officials – to find space to trade. Since, they do not trade in 'legal space' they are considered encroachers by the urban authorities, their wares and *laris* are often confiscated and they are prosecuted time and again. Many vendors feel victimised, but helpless, except where protected by a champion such as SEWA.

Main legislation affecting street vendors

Hawking and vending have been an integral part of Indian trade for centuries, and are now an integral part of the urban informal economy. Colonial institutions were superimposed over Indian reality, and many laws and legal principles remain fundamentally unchanged since independence, remaining hostile to street vendors. The Street Vendors (Protection of Livelihood and Regulation of Street Vending) Act, 2014 (GoI, 2014), is too new to have had much impact. Meanwhile, many restrictive byelaws and regulations remain, creating an architecture of regulation which effectively makes vending illegal and formalisation impossible. The next section reviews the main laws, rules and regulations in Ahmedabad pertaining to street trade, including local government, policing, highways and town planning legislation, demonstrating the complexity of the legal framework within which street vending operates.

The Bombay Provincial Municipal Corporation Act, 1949 (BPMC Act 1949)

Trade in Ahmedabad is regulated by AMC under the provisions of Bombay Provincial Municipal Corporation Act, 1949 (BPMC Act). The Act identifies the Municipal Corporation, Municipal Commissioner and Standing Committee as responsible for carrying out the provisions of the Act. It also describes the powers and duties of municipal authorities, granting municipal corporations the responsibility for maintenance, operation and development of public utilities in the city. Municipal corporations have two types of functions: 1) obligatory and 2) discretionary. The relevant sections of the BPMC Act are summarised below.

s.209 Power to acquire premises for improvement of public streets

s.209 empowers municipal authorities to acquire premises for the purpose of street improvement such as widening, expanding, building a new street.

s.226 Prohibition of projections upon streets

s.229 Prohibition of structure or fixtures which cause obstruction in streets

ss.226 and 229 prohibit the erection of any structure or stall on the streets that will obstruct passage of the public, or impede a drain or open channel. Such a structure may be removed by the Municipal Commissioner and the person responsible for the creation of the structure must incur the expense of removal.

s.231 Commissioner may, without notice, remove anything erected, deposited or hawked or exposed for sale in contravention of Act

s.234 Commissioner may permit booths, etc. to be erected on streets on festivals

s.231 allows removal of permanent or temporary structures in streets, and goods being hawked or sold in public places, and s.234 allows for the erection of street stalls during festivals.

s.328 Provision of new municipal markets and slaughter-houses

s.330 Prohibition of sale of commodities sold in municipal markets

Under s.328 the Commissioner is responsible for the provision and maintenance of municipal markets; under s.330 the Commissioner may prohibit sale of similar goods within a distance of 50 yards from a municipal market.

s.331 Opening of private markets and of private slaughter-houses

s.378 Private markets not to be kept open without licence

s.377 Prohibition of sale in municipal markets without licence of Commissioner

s.379 Prohibition of sale in unauthorised private markets

ss.331, 377, 378 and 379 set out powers for the approval and licensing of private markets, and restrictions on sale from unlicensed markets. Under s.377 the Commissioner should ensure that all traders in municipal markets are licensed.

s.384 Licences for sale in public places

s.384 establishes the need to obtain a licence from the Municipal Commissioner for hawking in a public place and confiscation of goods without prior notice in case of non-compliance. The Bombay Shops and Establishments Act, 1948, may prescribe the timings of any trade. Any person contravening the provisions shall be liable to have his goods seized.

s.431 Complaint concerning nuisances

Under ss.431 and 466, any person living in the city can register complaints about a nuisance in the city, and the Commissioner has powers to prevent such nuisance. Hawkers and vendors are often considered a nuisance by municipal agencies, urban elites and residents' associations.

s.466 Making of standing orders by Commissioner

s.466 covers the operations of a market to prevent a nuisance, fix trading times, prevent closure of shops and stalls, maintain cleanliness, require provision of ventilation and water and ensure circulation space.

Indian Penal Code, 1860

Only one section of the Indian Penal Code (IPC), is applicable to street vendors.

s.283. Danger or obstruction in public way or line of navigation

s.283 of the code allows a fine for anyone who causes danger, obstruction or injury in a public way. Nevertheless, this section is frequently invoked as the vendors are seen to be obstructing traffic and thus obstructing a public way – it is noted that if vehicles are parked on the footpath, these are not considered an obstruction!

Bombay Police Act, 1951

Hawkers are often evicted under s.67 and s.102 of the Bombay Police Act, 1951, which stipulates that anyone preventing smooth flow of traffic can be arrested and removed, and their tools and goods confiscated. According to the act, even a senior police officer cannot permit anyone to carry out vending on the streets.

s.67 Police to regulate traffic, etc., in streets

Under s.67 it is the duty of a police officer to regulate and control traffic on the streets, prevent obstruction, keep order in the street and other public places, and to regulate access to such places to prevent overcrowding.

s.102 Causing any obstruction in a street

s.102 requires that no person shall cause an obstruction by allowing any vehicle being loaded or unloaded to remain longer than necessary in a place, or by leaving in place any 'box, bale, package or other thing', or by setting out for sale any 'stall, booth, board, casket' etc.

Motor Vehicle Act, 1988

Section 201 of the Motor Vehicle Act, 1988, also penalises anyone who obstructs the flow of traffic on the public highway.

s.201 Penalty for causing obstruction to free flow of traffic

s.201 is designed to prohibit parking offences, specifying that whoever keeps a disabled (parked) vehicle in a place where it impedes the free flow of traffic will be liable to penalties. The wording can be applied to both motorised and non-motorised vehicles such as *laris* (hand carts) often used by vendors.

Criminal Procedure Code, 1973

s.151 Arrest to prevent the commission of cognisable offence

s.151 allows a police officer to arrest anyone about to commit an offence without orders from a magistrate and without a warrant, and the person can be detained in police custody for up to 24 hours. This is used in conjunction with the IPC or Police Act, to detain the street vendors if the local authority anticipates resistance to evictions.

Gujarat Town Planning and Urban Development Act, 1976 (GTPUD Act 1976)

In Gujarat, the development of urban areas, and preparation of development plans (DP) and town planning schemes (TPS), are determined by the Gujarat Town Planning and Urban Development Act, 1976 (GTPUD Act). The TPS for any local area allows land pooling and readjustment in areas of complex ownership. The planning authority replans or readjusts boundaries for urban development, but retains a proportion of each replanned private plot to provide infrastructure, social facilities, low-cost housing, or for commercial sale to raise funds for infra-structure development.

Chapter II, Development Area and Constitution of Area Development Authorities

s.12 describes the contents of the draft plan. It specifies reservation of land for public purposes such as schools colleges, medical and health institutes, cultural institutes, community facilities, but does not mention space requirements for the street vendors.

Chapter IV, Control of Development and Use of Land in Development Plans

Under s.35 vendors can be penalised if they carry out business in any area without permission, s.36 and s.37 permit removal of unauthorised temporary develop-ment by the Commissioner of Police and the District Magistrate.

Chapter V, Town Planning Schemes

The chapter explains the content of TPSs. s.40 in this chapter provides for the reservation of land within a TPS for the purposes of roads, open spaces, gardens, recreational activities etc. Under this section vendor markets could also be allo-cated. Draft plans have to be finalised within nine months of submission to the state government.

Licensing procedures

Under ss.376 and 337 of the BPMC Act, 1949, street vendors and hawkers have to obtain licences from AMC, issued by the Estates Department. Those trading food items need an additional licence from the Health Department. Licences are quite restrictive and should specify the period, restrictions, conditions, date of renewal etc. and the licensee must produce the licence required. Hawkers are not allowed to stay in one place for more than 30 minutes, and have to stand in a row. Hawkers have to use a *lari,* and should not 'disturb the streets and the footpaths by restricting the pathways' or use any 'sirens, horns or bells to grab the attention of the passer by'. The licence number and card has to be pasted to the front of the *lari* (CPPR, 2008).

Although, according to the BPMC Act, 1949, municipalities have to provide licences for vending and hawking, in reality the authorities are reluctant to issue licences, application forms are complicated and the criteria for acquiring a licence are onerous. The eligibility requirements for a vendors' licence include a ration card, proof of address, or voter ID card, and can act as a hurdle as many of the vendors and hawkers live in slums without a recognised address. In Mumbai there are 200,000 hawkers, but only 14,000 have been granted licences – comparable figures for Ahmedabad are not known. Women hawkers in the city are often subject to harassment by the police and municipal authorities as few women vendors in the city possess licences.

Street vendor's policy and legislation

The national street vendor policy was the culmination of many years of campaigning by NSAVI, leading to the NPUSV and its amendment in 2009. However, as a federal policy it had to be implemented by states, but take-up was slow.

After the widespread evictions of street vendors in preparation for the 2010 Delhi Commonwealth Games and continued country-wide struggles of street vendors, NASVI campaigned for the introduction of federal legislation and, proposed by the Minister of Housing and Urban Poverty Alleviation, a bill was introduced to parliament in 2012 and eventually passed as the Street Vendors (Protection of Livelihood and Regulation of Street Vending) Act, 2014 (GoI, 2014). In a complete reversal of previous repressive approaches, the legislation aimed at providing social security and livelihood rights to street vendors. It sets out procedures for regulating street vending, including the establishment of a town vending committee with street vendor representation; issuing a certificate of vending; the rights and obligations of street vendors; designation of vending zones; processes for relocation and evictions; and prevention of harassment of street vendors. Crucially the law defines the concept of a 'natural market' as: 'a market where sellers and buyers have traditionally congregated for the sale and purchase of products or services and has been determined as such by the local authority on the recommendations of the Town Vending Committee' (s.2.e).

In Ahmedabad the policy was implemented before the Act was passed, through a complex legal process known as the Street Vendor's Policy for Ahmedabad, which adds another layer of complexity to the management of street trade in the city, as discussed further below.

The impact of legislation

It is clear from this review of legislation that street vendors exist within a harsh regulatory framework in which the protection of their rights is weak. The law brands vendors and hawkers as a civic nuisance and encourages authorities to evict them. Vendors have to deal with multiple agencies – the AMC and its various departments, the police and the planning authorities – leaving them open to exploitation and extortion. The 2014 Street Vendors Act is a fundamental change in approach, but has yet to be implemented in many Indian states.

In the meantime, 14 sections of the BPMC Act, 1949, regulate vending and hawking in Ahmedabad. The provisions are supplemented by the Bombay Shops and Establishments Act, 1948; together they regulate the use of pavements. Vendors and hawkers can be penalised and/or harassed at any time through the application of sections of the Act. The municipal laws do not directly prohibit vending and hawking as a profession, but impose a gamut of restrictions on it.

Any of the legislation discussed above can be used to evict the street vendors. For three days, from 17 December 2011, s.283 of the IPC was used by the Ahmedabad police to evict street vendors from Bhadra market in the historic city centre, for urban improvement works. Field visits and news reports suggest that at least 50 street vendors were prosecuted using this Act; the decision to evict appeared to be taken by a new Deputy Commissioner of Police (DNA, 2011). After the work was complete, having no other livelihood, street vendors gradually filtered back into the space.

Ss.231 and 384 of the BPMC Act, 1949, are the sections most frequently used to evict and prosecute street vendors, but the Acts need to be amended in order to remove the anomaly between a legal vendor and illegal obstruction. In fact, these are *ultra vires* by virtue of Article 39 (a) and (b): Article 14: Article 19 (1) (g) and Article 21 of the Constitution. Similarly, the Union Government should amend s.283 of the Indian Penal Code, 1860, to make an exception, so that vendors and hawkers are not prosecuted for occupying a public way.

As Bhowmik (2001, 2003) has argued, regularisation would imply that hawkers would not be forced to bribe the authorities in order to trade, and municipal authorities would increase their revenues through fees collected from hawkers. There is considerable potential for improved design of roads and pavements such that the street vendors can continue their trade without inconveniencing other road users. There has to be a lawfully accepted resolution to protect vendors' business interests to create conditions such that they are not constantly harassed by enforcement authorities. It can also be argued that the solution to such conflicts lies not in the law but in planning. TPSs can provide an enabling mechanism for reserving lands for market, but the GTPUD Act, 1976, does not recognise 'natural markets' and while urban plans allocate space for public amenities such as parks, hospitals, or community space, they do not take into account places that can be developed into 'natural markets' for hawkers – for example, at rail or bus termini etc. (Bhowmik, 2001). Master plans do not allocate space to vendors as they follow westernised concepts, ignoring the Indian traditional forms.

Meanwhile, 'weekly markets' struggle to survive and 'natural markets' are altogether ignored.

In relation to urban planning, changes to regulations would be required, so that master plans and local area plans cater to the needs of space for street vending. Suitable spatial planning 'norms' for reservation of space for street vendors in accordance with their current population and projected growth need to be devised. State governments have been advised to 'remove the restrictive provisions in the Municipal Acts to make street vendors inclusive in the city plan/cityscape' (Bhowmik, 2003). Interestingly, the TPS provision has been implemented in Rajkot City, Gujarat, in an innovative scheme wherein a commercial plot has been allocated for the development of a vendors' market in a prime location in the city. The vendors occupy the market after 4pm every day. Rajkot Municipal Corporation has installed lights and provided toilet facilities.

Claiming rights through PIL

The section outlines how, under the guidance of SEWA, the vendors of Ahmedabad took recourse to the Courts, to challenge the persistent, unpredictable and often violent evictions faced by their members. Supported by *pro bono* legal advice, SEWA used PIL to launch a four-year legal battle to win security for the vendors. The use of constitutional provisions to attain justice for the urban poor through PIL is specific to India, and PILs are often used as *de facto* instruments in setting public policies.

The Constitution of India establishes Fundamental Rights (Part II) and Directive Principles of State Policy (Part IV). The Fundamental Rights are non-negotiable, basic principles of the functioning of the Indian state. These include Article 21, the Right to Life, under which, right to livelihood and right to shelter are subsumed; Article 14, the Right to Equality before Law; and Article 19 (1) (g) the Right to Practice Profession and Trade. All other laws have to abide by the articles of the Fundamental Rights, but they are not directly enforceable through a court of law.

Article 32 of the Constitution gives citizens the right to petition the Supreme Court in matters that violate their Fundamental Rights, and Article 226 gives similar rights to approach the state High Court. Citizens can approach the Higher Courts in public interest if the fundamental rights of a group of people are violated by action of the state.

In 1980s, the definition of who could initiate PIL was widened to include any concerned citizen representing a public interest, including third parties. This became an important mechanism for human rights groups seeking justice for deprived populations. PILs have since been used to challenge non-implementation of existing legislation, raise public awareness and public policy debates and to force the government to enact new legislation.

On 4 September 2006, a PIL was filed by SEWA, with the plea that AMC be required to implement the draft NPUSV and restricted from evicting street vendors, which violates Article 19 (1) (g) of the Constitution, and that AMC and

Ahmedabad Urban Development Authority (AUDA) be required to provide spaces for street vending in TPSs. Only the first element of the plea was admitted by the High Court.

The PIL went through many twists and turns, as summarised below. The PIL was admitted and High Court notices were issued to the respondents, AMC, AUDA and the state government. The High Court order of 2 November 2006 directed AMC to state the steps taken with regards to the vendors' issues to date, and asked AMC to inform the High Court its policy on rehabilitation of street vendors.

AMC replied, stating that it had a Scheme of 1988, but SEWA replied that the city had expanded since 1988 and harassment of street vendors continued, which required a new scheme to be prepared according to NPUSV. SEWA further argued that earlier representations to AMC had produced no response as there is no mechanism for poor populations to negotiate with local government on matters of their concern.

On 29 November 2006, the Gujarat High Court ordered AMC and AUDA to have a meeting with SEWA to sort out the grievances and requested AMC not to disturb any vendors until the grievances were addressed. The High Court had therefore indicated that policy-level disputes could be negotiated at the local level under court direction, where an organisation representing the affected population existed. However, AMC filed a counter affidavit on 23 January 2007 that it had already passed the 1988 Scheme.

Following a delay of nearly two years, on 17 November 2008, the High Court, concerned about lack of progress, ordered AMC to finalise the scheme allocating vending spaces within three months from receiving proposals from SEWA. On 18 December 2008, SEWA filed an affidavit stating that they had submitted proposals at a meeting with AMC, but the latter had not prepared a scheme. Instead, SEWA had prepared a scheme for five of the nine plots identified by AMC for vendors' markets – the other four plots were not viable as trading sites. AMC did not comply with the High Court order.

Finally, on 16 March 2009, the High Court again ordered AMC to prepare a scheme within one year and place it before the Court on or before 6 March 2010. The AMC commissioned consultants who prepared the Street Vendors Policy for Ahmedabad City, which was submitted to the Court. The Court did not accept the policy as a scheme and required the Municipal Commissioner to appear in person to commit to preparing a scheme, which was finally prepared and placed before the Court on 2 March 2010, three and half years after filing the original PIL.

SEWA had concerns about the scheme, which was based on limited survey and completely overlooked the concept of 'natural markets', and had not even mapped the existing markets. Instead, the scheme placed before the High Court proposed three categories of vending spaces:

1) *green vending zone* on 15m roads, unrestricted vending;
2) *amber vending zone* on 15–30m roads, restricted vending from 6–9am and 6–9pm with permission from AMC and

3) *red vending zone* on 30m roads and within 200m of a heritage monument, no vending.

SEWA raised objections to the proposed scheme and, on 16 March 2010, the High Court ordered that if SEWA was not satisfied with the scheme and considered that it contained inadequate safeguards for the vendors, both organisations should meet to find solutions and modify the scheme. Again, the emphasis was on negotiation. SEWA submitted the map of 174 'natural markets' in the city, asking AMC to regularise them and either prepare maps for development or relocate those that could not be regularised. AMC instead accepted another of SEWA's suggestion to form a town vending committee (TVC), also recommended in the NPUSV, with significant vendor representation.

The High Court, ultimately accepted the street vendor scheme prepared by AMC. SEWA in its affidavit of 25 August 2010, stated that if implemented the proposed scheme would displace 129 out of 174 natural markets in Ahmedabad, depriving 38,908 vendors of their livelihoods and 946,015 customers of local services. In the final judgment of 27 August 2010, the High Court, however, accepted the AMC scheme, after taking suggestions from SEWA, requiring it to be implemented within six months. The High Court also stated that AMC could proceed with eviction of unauthorised vendors according to the scheme after notifying them through a public address system.

The Ahmedabad Street Vendor's Scheme, 2010, now determines the functions and composition of the Ahmedabad TVC, and classifies major streets in the city by the level of street vending permitted, determined by the red, amber and green zones. There are two crucial problems with this approach: first, that it ignores 'natural markets' and, second, that it will lead to extensive evictions.

Thus, powers of policy-making were transferred to the courts, and the long-fought PIL brought mixed success for the vendors, as, although the needs of vendors were recognised and a TVC set up, the scheme would be very restrictive if fully implemented. However, the scheme has become so controversial that it has not been published and thus not implemented, but street vendor displacements have continued for infrastructure projects and along roads designated as 'Model Roads'. The negotiations between SEWA and AMC continue through formal and personal channels. The Bhadra Fort area, with about 3,500 vendors, has been redeveloped through JNNURM funding and caused massive displacements, although after three years many vendors have now invaded the newly paved space. Despite the positive role played by the Gujarat High Court, the PIL outcome was highly unpredictable, and since the scheme is now accepted by the High Court it can eventually be published and used to evict the street vendors.

Conclusions

The municipal and police laws that impose restrictions on street vending do not directly prohibit street vending as a profession, but impose stringent restrictions on the use of urban space for street vending, making it virtually impossible to trade legally. The Acts are archaic and fail to meet the challenges posed by the current urban context, particularly in relation to the changing economy, increase in self-employment, need for jobs for the city's young workforce and saturation of the formal sector.

In most of the legislation vendors are viewed as a problem to be controlled, or as a nuisance or obstruction, rather than as enterprises that contribute to the urban economy. Since vendors typically lack legal status and recognition, they frequently experience harassment and evictions by enforcement agencies. Although the sector provides employment, it tends to be perceived as antisocial, anti-developmental, dirty, unattractive and unhygienic, and some businesses, such as food stalls, face additional checks by the Food and Drug Administration. Negative attitudes and neglect have meant that the sector is ignored in town planning regulations and treated as unplanned urban activity. The reality, however, is that the sector comprises self-employed people who are trying to earn a legitimate living and are an integral part of the city's economy who ask only *'do tokriki-jagah'* (space for two baskets, i.e. a life of dignity).

For street vending, an ambiguity within the Constitution lies in the fact that Article 19 (1) (g) relating to freedom to practice professions and trades does not clarify whether the occupation of hawkers and vendors falls within its scope. Furthermore, while interpreting the term 'public interest' in Article 19, it is difficult to argue that the activities of the hawkers are in the public interest. Indeed, several laws, including police and town planning Acts, consider trading on the road as an obstruction and hence a nuisance that needs to be removed in the 'public interest'. Thus, although, the Constitution guarantees freedom of trade and a right to a livelihood, other legislation contravenes this provision. Hence, street vendors are constantly under the threat of displacement and at best offered alternate sites for vending. While the new Street Vendors (Protection of Livelihood and Regulation of Street Vending) Act, 2014, represents recognition for this vulnerable group, its focus on 'natural markets' restricts vending elsewhere.

The judicial activism of SEWA in pursuing the long-fought PIL to implement the NPUSV was based on respect for the rule of formal law evident in India. However, this effectively transferred policy-making from the local authority to the courts, with outcomes relying on the awareness and wisdom of the presiding judge with limited scope of challenge. For SEWA and Ahmedabad's vendors the outcome was mixed. Ahmedabad now has a vending scheme, but its operation could be extremely repressive. The 2014 Street Vendor Act is an innovation in that it aims to strengthen social protection and livelihood rights for street vendors, but until it is fully implemented by states and while other conflicting legislation remains in force, the potential for evictions remains a constant threat.

In academic literature, there is mixed opinion on the extent to which the PIL process benefits the urban poor. In fact, Ghertner (2008), Bhan (2011), Dupont and Ramanathan (2006, 2008) indicate that PILs emerging from middle-class activism have been more successful than those seeking to support the urban poor (Srivastava, 2009; Kundu, 2009). The PIL brought by SEWA to protect street vendors in Ahmedabad has brought their plight to the fore but has been subverted in its implementation. Nevertheless, establishing formal legal rights and a mechanism for judicial representation and claim, as India's street vendors have achieved, is a crucial step in increasing rights and reducing vulnerability in practice.

Acknowledgements

Many people contributed to the research; our particular thanks to Manali Shah, Shalini Trivedi and Geeta Goshti of SEWA.

Note

1   Ahmedabad's Utarayan kite festival in mid-January is a two-day holiday marked by widespread celebration in the city.

References

Bhan, G. (2011) 'This is no longer the city I once knew'. Evictions, the urban poor and right to the city in millennial Delhi, *Environment and Urbanization*, 21(1): 127–42.

Bhowmik, S. (2001) Hawkers in the urban informal sector: a study of street vendors in seven cities, Delhi: National Association of Street Vendors in India (NASVI), http://wiego.org/publications/hawkers-and-urban-informal-sector-study-street-vending-seven-cities, accessed January 2015.

Bhowmik, S. (2003) Urban responses to street trading: India, Research paper, *Urban Research Symposium*, Washington DC: World Bank, http://wiego.org/publications/urban-responses-street-trading-india, accessed January 2015.

CCT (2002) Crime against humanity – an inquiry into the carnage in Gujarat, Volume I: List of Incidents and Evidence, Mumbai: Citizens for Peace and Justice, Concerned Citizens Tribunal (CCT), http://www.sabrang.com/tribunal/, accessed January 2015.

CPPR (2008) Rules and regulations governing the licensing of vegetable vendors in Ahmedabad, (CPPR) Centre for Public Policy Research, http://livelihoodfreedom.in/L%203%20FOR%20SITE/Ahmedabad/VEGETABLE%20SELLERS.pdf, accessed April 2011.

Dalwadi, S. (2003) Mapping of natural markets in Ahmedabad, Unpublished research report, Ahmedabad: SEWA (Self-Employed Women's Association).

Debroy, B. and Bhandari, L. (2011) State of economic freedom in India, in Ankalesaria Aiyar, S.S. (ed.), *Economic Freedom of the States of India, 2011*, New Delhi: CATO Institute: 17–32, http://www.cato.org/economic-freedom-states-india, accessed January 2015.

DNA (2011) Lal Darwaza hawkers and DCP at loggerheads, in *Daily News Analysis (DNA)*, 19 December: 3, http://www.highbeam.com/doc/1P3-2542128011.html, accessed January 2015.

Dupont, V. and Ramanathan, U. (2008) The Courts and squatter settlements in Delhi or the interventions of the judiciary in urban governance, in Baud, I. and de Wit, J. (eds), *New Forms of Urban Governance in India: Models, Networks and Contestations*, Delhi: Sage: 312–44.

Ghertner, D.A. (2008) Analysis of new legal discourse behind Delhi's slum demolitions, *Economic and Political Weekly*, 43(20): 57–66.

GoI (2014) Street Vendors (Protection of Livelihood and Regulation of Street Vending) Act, 2014, http://www.indiacode.nic.in/acts2014/7%20of%202014.pdf, accessed January 2015.

Hirway, I. and Mahadevia, D. (2004) *The Gujarat Human Development Report, 2004*, http://www.in.undp.org/content/india/en/home/library/hdr/human-development-reports/State_Human_Development_Reports/Gujarat.html, accessed January 2015.

India, Planning Commission (2012) *Press Note on Poverty Estimates, 2009–10*, March, New Delhi: Government of India, http://planningcommission.nic.in/news/press_pov1903.pdf, accessed April 2016.

Jajoo, K. (2012) Measuring the street traders contribution to the urban economy (examining Ahmedabad) [Thesis], CEPT University Library, Ahmedabad.

Kundu, D. (2009) Elite capture and marginalization of the poor in participatory urban governance: a case of resident welfare associations in metro cities, in MoHUPA and UNDP (eds), *India: Urban Poverty Report 2009*, New Delhi: Oxford University Press: 271–86.

Mahadevia, D. (2007) A city with many borders: beyond ghettoization in Ahmedabad, in Shaw, A. (ed.), *Indian Cities in Transition*, Hyderabad: Orient Longman: 315–40.

Mahadevia, D. (2010) Urban reforms in three cities: Bangalore, Ahmedabad, and Patna, in Chand, V. (ed.), *Public Service Delivery in India: Understanding the Reform Process*, New Delhi: Oxford University Press: 226–95.

Mahadevia, D. (2012) Decent work in Ahmedabad: an integrated approach, ILO Asia-Pacific Working Paper Series, Bangkok: ILO (International Labour Organization, Regional Office for Asia and the Pacific), http://www.ilo.org/wcmsp5/groups/public/---asia/---ro-bangkok/documents/publication/wcms_181745.pdf, accessed January 2015

Mahadevia, D. (2013) Social security for the urban poor – a study in Gujarat, in Kannan, K.P. and Breman. J. (eds), *The Long Road to Social Security: Assessing the Implementation of National Social Security Initiatives for the Working Poor in India*, New Delhi: Oxford University Press: 335–70.

MoHA (2011) Provisional population totals paper, Office of Registrar General and Census Commissioner, MoHA (Ministry of Home Affairs), Delhi, http://censusindia.gov.in/2011-prov-results/paper2/prov_results_paper2_indiavol2.html, accessed 15 October 2011.

MHUPA (2009) National policy on urban street vendors, 2009, New Delhi: MHUPA (Ministry of Housing and Urban Poverty Alleviation), http://mhupa.gov.in/Default.aspx?ReturnUrl=%2fw_new%2fStreetVendorsBill.pdf, accessed January 2015.

OIA (2010) Report of public hearing on habitat and livelihood displacements, Ahmedabad: OIA (Our Inclusive Ahmedabad), http://cept.ac.in/UserFiles/File/CUE/Advocacy/Public%20Hearing%20on%20Displacements.pdf, accessed January 2015.

Ramanathan, U. (2006) Illegality and urban poor, *Economic and Political Weekly*, 41(29): 3193–97.

Srivastava, S. (2009) Urban spaces, Disney-divinity and moral middle classes in Delhi, *Economic and Political Weekly*, 44(26 and 27): 338–45.

9　Law and litigation in street trader livelihoods

Durban, South Africa

Caroline Skinner

Summary

Durban was, for many years, considered a model of inclusive planning for street traders, but closer scrutiny shows a more mixed picture. This chapter critically assesses the role of law and litigation in state approaches towards street traders in Durban. The chapter tracks policy shifts from the early apartheid years, through transition, to the 1990s enabling approach of Warwick Junction and the Durban Informal Economy Policy, and the post-2004 repressive pro-development agenda. It then explores how traders have used litigation to fight development proposals for central market space, and to claim constitutional rights, arguing that hard-won gains must continually be defended.

Introduction

This chapter outlines ebbs and flows of inclusion and exclusion of street traders in Durban, demonstrating how, in a city that was for many years considered a model of inclusive planning for street traders (ILO, 2002; UN-Habitat, 2006), the context could rapidly change, and exploring how street traders have used law and litigation to fight back. The chapter draws on many years of research and advocacy work by the author, building on and updating previous work (Skinner, 2008). It is informed by the increasing focus on the role of law and legal reforms as a means to address poverty (Brown and Rakodi, 2006). This was given particular impetus with United Nations Development Programme's Commission for Legal Empowerment for the Poor (CLEP). Legal empowerment they define as follows:

> The process through which the poor become protected and are enabled to use the law to advance their rights and their interests, vis-à-vis the state and in the market. It involves the poor realising their full rights, and reaping the opportunities that flow from that, through public support and their own efforts as well as the efforts of their supporters and wider networks.
>
> UNDP, 2008a: 39

The final report pays considerable attention to the informal economy. It identifies 'the right to work' with specific reference to 'the right to vend and the right to a work space' as 'basic commercial rights' (UNDP, 2008b: 220–24). It outlines key legal mechanisms to empower informal businesses like simplified licensing and permit procedures; bylaws that allow street traders to operate in public space; legal frameworks that enshrine economic rights (like access to finance and markets); and includes references to use rights to urban public land. It even identifies legal rights of association and representation in policy-making institutions (ibid.: 223). The Commission focused attention on the specific legal needs of street traders providing an agenda for change. The report notes that legal empowerment is 'a country and context-based approach that takes place at both the national and local levels'. Interrogating a case which is has been considered a 'good practice' allows an assessment of the extent of legal empowerment deficits, but also potential shortfalls in the overall approach.

Size and contribution of the street trading economy

According to the April–June 2015 Quarterly Labour Force Survey statistics, 2,661,000 South Africans work in the informal sector (Statistics South Africa, 2015: iv),[1] constituting 17% of total employment in the country. Retail has long been the dominant activity in the South African informal sector. In the April–June 2015 survey of those reporting to work in informal enterprises, 1,067,000, or 40%, reported working in wholesale or retail trade. Although individual incomes are low, cumulatively these activities contribute significantly to the economy. Statistics South Africa estimates that the informal sector contributes 5% to gross national product (GNP) (Fourie and Kerr, 2015). Ligthelm (2006) estimated the share of informal trade sector at approximately R32 billion[2] constituting 10% of retail trade. The comparable figures for South Africa's two largest chain stores, Pick n Pay and Shoprite, were R32 billion and R27 billion, respectively (Ligthelm and Lamb, 2004).

There have been three censuses/surveys of street traders in Durban that give an insight into the overall numbers and nature of street trade in the city. A 1997 census estimated there were 19,301 street traders operating in the Durban area, over 10,000 of whom operated in the inner city (Data Research Africa, 1998). A second census conducted in 2009 highlighted the extent to which numbers of traders fluctuate at different times of the week and month, but estimated there were just under 50,000 traders operating in the metro area (Reform Development Consulting, 2010). This suggests that the numbers of street traders have more than doubled over the 12-year period between counts.

For a long period, street traders in Durban were largely women. Both the 1997 census and a 2003 survey of just over 4,700 Durban street traders (KMT Cultural Enterprises, 2003) found 59% of traders were women. The 2009 survey of 4,000 traders attached to the census, however, found that only 44% of those interviewed were women (StreetNet, 2011:1). This is in line with national trends in employment in the informal sector. Although difficult to quantify, the street trading economy is clearly a source of employment for foreign migrants (Hunter and Skinner, 2003). The findings on dependency ratios suggest that street trading is an important activity in terms of household well-being in the city. The 2003 survey, for example, found that 88% of traders were the sole breadwinners in their families and over three-quarters of traders reported that they had three or more dependents, with over 30% of traders reporting seven or more dependents (Hunter and Skinner, 2003: 12). In the 2009 survey there were, on average, 4.3 individuals dependent on one street trader's income. All three surveys found, the majority of traders were selling food. There is mounting evidence on the important role the informal economy is playing in food security for low-income households (Crush and Frayne 2011), and the African Food Security Urban Network (AFSUN) reports (AFSUN, 2016).

Apartheid repression to regulated trading: the early years

Evidence suggests that processes of exclusion, followed by limited inclusion of street traders in the apartheid period and during the transition, were partly determined by a complex interplay of national and local government policy imperatives, mediated by private sector interests and traders responses, outlined below.

Trader repression under apartheid

Durban street traders have been penalised through harsh regulations for over a century. However, Nesvag (2000: 35–6) argues that until the 1930s street trade was not as harshly regulated as it would be in the period to come, noting that local native policy throughout this period, while repressive, was only partially effective, as it was often contradictory and full of loopholes, due to deep divisions of interest between the various actors in white political and business circles in Durban. These loopholes were largely closed with the implementation of apartheid legislation. The two most significant pieces of legislation affecting the development and regulation of informal activity were the Group Areas Act (1950) and the Black (Urban Areas) Consolidation Act (No. 25 of 1945). The former disallowed black South Africans from accessing the more viable trading or manufacturing points and the latter imposed restrictions on economic activities even in so-called 'black areas'.

At a local level the Durban City street trading bylaws (1962) outlawed street trading in the city. A scan of local newspapers of the 1960s and early 1970s indicates that traders were severely harassed. For example, in 1966 the *Daily News* reported that '485 people have been charged in less than six months with illegal street trading' (17 October 1966) and in 1973 it described police as 'fighting a battle' against illegal hawkers and quoted Durban's Chief Licensing Officer as saying that 'in terms of Government legislation no Africans are allowed to trade in white urban areas unless they are employees of a white person' (28 May 1973).

In 1973 the Natal Ordinance (11/1973) was introduced, which allowed very limited trading which was regulated by what became known as the 'move on' laws. This provincial legislation restricted hawking of goods within 100m of a fixed formal business. It prevented hawkers from taking up fixed stands by allowing them to occupy a spot for only 15 minutes, after which they were to move at least 25m away. No sales point could be occupied more than once on the same day. The harassment of traders did not abate significantly. For example, in 1977 the *Daily News* reported that more than 70 fresh produce hawkers trading in Warwick Avenue were given less than 24 hours 'to clear the area permanently' or face 'drastic police action' (25 January 1977). The police justified their actions on the grounds of a perceived link between crime and trading, and their concern for consumers' health.

Reflecting on this period, Rogerson and Hart (1989: 32) argue that South African urban authorities 'fashioned and refined some of the most sophisticated sets of anti-street trader measures anywhere in the developing world'. They point out that until the early 1980s hawkers in South Africa were subject to 'a well-entrenched tradition of repression, persecution and prosecution'. This was underpinned by both national and local legislation.

Transition from apartheid: regulated trading

In the early 1980s, the Progressive Federal Party won the local government elections in the city of Durban. The new local council, more liberal than its predecessor, the National Party, reviewed the city's approach to street trading. The *Daily News*, for example, stated that:

> the city council recognised the need to make allowance for the economic needs of at least some of the more than 100,000 people flocking to the peripheries of the city every year in the hope of finding work in a shrinking urban job market' and added that a report commissioned by the city on the issue – the Hawker Report – demonstrated that 'hawkers were a fact of life throughout the city.
>
> 18 June 1987

In 1985 a subcommittee of councillors was formed to find practical ways of implementing more favourable policies for street traders. They decided to introduce a hawker's licence which allowed informal trading to take place. This ushered in a new era of regulated trading.

This new approach coincided with developments at a national level. From the mid-1980s, influx control laws became increasingly unenforceable and were abolished in 1986, and in 1987 the national government gazetted the National White Paper on Privatisation and Deregulation, which introduced a more tolerant approach to black small business in general. The national change of attitude to informal activities culminated in the 1991 Businesses Act (RoS, 1991). The Act was the key measure for deregulation of business activities, removing barriers to the operation of informal activities. The Act specified that local authorities, with a few exceptions, have to allow immobile informal trading and cannot restrict traders to specified hours, places, goods or services. The Act thus stopped conservative municipalities from disallowing trading, fundamentally changing the environment in which street traders operated. This formed part of a broader strategy of free enterprise promotion of nationalist government in the dying days of apartheid. As a result of the Act, the number of street traders dramatically increased in urban centres across the country. Local authorities, however, complained that they were unable to cope, which led, in 1993, to the Amended Businesses Act. The amendment allowed local authorities to formulate street trading bylaws, outlining what local authorities can and cannot include. The amendment also allowed local authorities to declare restricted and prohibited trade zones.

In line with Amended Businesses Act, Durban promulgated new street trader bylaws in 1995. The bylaws ushered in a new regime of re-regulation that continues today. Although Durban City Council did declare certain areas of the inner city prohibited trade zones, most of it was declared a restricted trade zone and sites were demarcated. Although the Durban City Council's actions did reduce the number of traders, the majority were accommodated. This is thus a case where national legislation facilitated the accommodation of traders on the street. However, street trader bylaws in Durban, as is the case in other cities, informed by the Amended Businesses Act, state that traders who contravene key elements of the bylaws will have their goods removed and impounded and be liable to pay a fine or be imprisoned. This means that, although these are livelihood activities, criminal rather than administrative sanctions apply. The tools for harsh management of these activities thus remain.

One of the requirements outlined in the Amended Businesses Act was that interested and affected parties be consulted. In the run-up to formulating Durban's street trading bylaws the City Council insisted that there be an umbrella body of trader organisations with which it could negotiate. The Informal Trade Management Board (ITMB) was thus established. Over time, the ITMB became more independent and by the late 1990s had support among street traders operating throughout the city. A few months prior to this the Self-Employed Women's Union (SEWU), an organisation modelled on the Self-Employed Women's Association (SEWA) in India, had been launched in Durban and was active, particularly among inner-city traders. Both of these organisations lobbied and negotiated with the Durban local authorities for the infrastructure needs of their members. During this period shelters were built for traders throughout the inner city. The activities of these organisations contributed to the pressure to incorporate traders into city plans. SEWU's sustained engagement, in particular, ensured that interventions were appropriate to women traders' needs (Devenish and Skinner, 2006).

Innovations: mid-1990s–2004

The Warwick Junction Project

In 1995 and 1996 the foundations were laid for what has become widely recognised as a good example of integrating street traders into city plans – the Warwick Junction urban renewal project (Dobson and Skinner, 2009). The Warwick Junction precinct lies on the edge of the inner city, contains a confluence of rail, taxi and bus transport and is the primary transport node for the inner city. By the mid-1990s it was estimated that an average of 300,000 commuters travelled through the precinct every day. Given the high pedestrian numbers, this area has always been a natural market for street traders. With deregulation, traders flocked to the area and, by the mid-1990s, it was estimated to house over 4,000 traders. At the time, however, there were real concerns about urban management of the area, and it was perceived to be a crime and grime hotspot.

The project was mandated to focus on safety, cleanliness, trading and employment opportunities and the efficiency of public transport, among other issues. The area-based team initiated substantial capital works and established a number of operations teams to deal with issues such as kerbside cleaning, ablution facilities, child-care facilities and people sleeping on the pavement. Between 1997, when the project was launched, and 2000, the project delivered in excess of R40 million in capital works. These included the upgrading of public transport facilities, and street lighting, landscaping and environmental improvements. Of particular relevance to street traders were the provision of street trading facilities, particularly shelters, the establishment of a dedicated market for the traditional medicine traders, and renovations to the fresh produce market building – the Early Morning Market (EMM). Early on in the project, an old warehouse located in the area was renovated to house the project office and serve as a community centre. This project has been lauded by many commentators (Grest, 2000; Saunders, 2004; Horn, 2004) and has won a number of awards.[3]

The first large infrastructure intervention was the development of the traditional medicine market. For many years Warwick Junction had been a hub of activity for traditional medicine or *muthi* trading, but no provisions were made for the *muthi* traders and, in the late 1980s and early 1990s, *muthi* traders were frequently seen sleeping on the streets at night under plastic sheets. Warwick Junction Project staff, in consultation with the existing traders, developed a facility for them in the Warwick Junction area, and established a dedicated built market, with shelter, storage, water and toilet facilities that accommodates 550 traders. These facilities have significantly improved their working environment. Writing for the *Sunday Times*, architect Silberman placed the traditional medicine market under the banner of 'The Best of the Century', stating this is 'one of the first South African structures which addresses – and celebrates – the informal traders who have come to dominate our city centres' (*Sunday Times*, 19 December 1999).

The traditional medicine market was one among a number of sector-specific interventions. By the mid-1990s there was vibrant trade in *mielies* (roasted corn on the cob) and the Zulu delicacy boiled ox heads, mostly sold by women. These are potentially hazardous trades as they are prepared on open fires but, rather than ban them, council officials worked with the cooks to design appropriate infrastructure. Another particularly precarious informal activity is collecting cardboard, also dominated by women. In the late 1990s, there were over 500 women collecting an estimated 30 tonnes of cardboard in the inner city every day, but being exploited by unscrupulous buyers. The project established an inner-city buy-back centre in Warwick so the cardboard could be sold directly to the recycling company. This increased the collectors' (albeit still low) incomes by 300%.

Much reference is made to the consultative nature of the project. The general secretary of SEWU at the time of the bulk of the capital works projects in Warwick Junction notes that 'the manner in which informal traders and other key stakeholders were engaged was qualitatively different from the type of consultation that is more often seen when project managers try to secure buy-in from stakeholders' (Horn, 2004: 211). She also states that 'the Council afforded informal traders the opportunity to participate on a sustained and continuous basis in negotiations about their needs and priorities ... in a low key way, often on an issue-by-issue basis' (ibid.: 211–2). Project staff worked with both trader organisations and individual traders. Informal traders were comparatively well organised, with both the SEWU and the ITMB active in the area.

Consultative forums encouraged and supported high levels of volunteerism in the ongoing market management. A good example of this is the initiative Traders Against Crime (TAC) group, formed by traders in 1997 in conjunction with the South African Police Service and the Durban Metropolitan Police. TAC members have been trained by the City Police and they patrol trading areas, alerting the authorities when action is needed. Since its inception the initiative has seen positive results, having substantially reduced both petty and more serious crime in the inner city. In the Warwick Junction area, for example, during the first 18 months of TAC's operation, there was only one murder. This was compared to the 50 murders in the previous year. The 300 committee members now operate in the most densely traded areas of the city.

The former project manager of the Warwick Junction Project described how the project 'worked with the dynamism' it found on the streets (interview, Richard Dobson, 28 February 2007). Rather than attempting to change the extent and nature of these activities, the project, by means of thorough consultation, engaged with the infrastructure and activity specific needs of street traders and responded accordingly. Keith Hart commented about the Warwick area that

'Durban has provided an exhilarating proof of how poor people, in sensitive collaboration with urban planners, can enliven a city centre, generate employment for themselves and expand services for the population at large' (interview, Keith Hart, Professor, Goldsmiths College, University of London, 29 May 2008).

While the Warwick Junction Project staff concentrated on this precinct, the Informal Trade Small Business Opportunities (ITSBO) Branch were responsible for managing and developing street trading in all other parts of the city. Trader shelters were constructed along many of Durban's main inner-city streets. Attractive shelters and trader storage facilities were built at the beachfront, where there is a predominance of craft sellers. Space has been allocated to between 350 and 700 traders (the number fluctuates seasonally). Further trader infrastructure was developed outside the inner city – in the former black townships of KwaMashu to the north and Umlazi to the south and in the former Indian townships areas of Phoenix and Chatsworth. In the inner city, the trader infrastructure doubtlessly improved traders' work environments, and interventions were often made in close collaboration with trader groupings. In outlying areas there is evidence that sometimes these developments were not as well conceived. For example, the Chatsworth market lay empty for a number of years since there was no allocations policy. There was also much controversy over the Phoenix market, which was badly located (Attwood, 2000).

The former president of SEWU and a longstanding traditional medicine trader, reflecting back on the period of integration of street traders, noted the importance of having a permit to trade, stating 'If you have the permit you can eat, you trade the way you want to trade, no one is disturbing you' (interview, Zodwa Khumalo, 31 March 2008). In effect the Warwick Junction Project built on 'the right to vend' and provided work space with basic infrastructure – in other words the 'basic commercial rights' outlined by the CLEP report. This was done through innovative participatory processes.

Institutionalising a progressive stance: Durban's Informal Economy Policy

In the late 1990s the city acknowledged that, although progress had been made with street trading, there was no overall policy guiding the city's interventions with respect to the informal economy. Interventions were thus *ad hoc*, with different departments responding in different ways. Further, it was acknowledged that little attention had been paid to informal workers other than street traders (eThekwini Unicity, 2001: 6). In November 1999 a technical task team was formed and mandated to formulate 'an effective and inclusive' informal economy policy. As outlined in Lund and Skinner's analysis (2004), the process was evidence based, drawing on specially commissioned research, and entailed substantial consultation with stakeholders both in and outside the council; the Informal Economy Policy was published in 2001.

The policy's point of departure was that the informal economy is critical to economic development. A number of specific aspects of this policy are worth highlighting. It makes suggestions for improving the management of informal economy activities with respect to registration, site allocation and charges for operating. The policy suggests a legislative review stating specifically that 'the bylaws should reflect the overall policy move away from sanction and control, towards support and the creation of new opportunities in a well-managed environment' (eThekwini Unicity, 2001: 20). It suggests that area-based management zones be established, since in the Warwick Junction case these had offered an opportunity to resolve coordination problems and encourage interest groups to participate in planning and management. It argues that decentralised management, combined with a programme that helps informal workers' representatives articulate their needs, was likely to create better work environments. It lays out a capacity-building programme for organisations and local government officials. However, it also commits the city to providing support services to people who work for very small enterprises, thus addressing a gap in the national government's small business policy in which the informal sector had and continues to be ignored (Rogerson, 2004; Devey *et al.*, 2006). It suggests the provision of basic business skills training, legal advice, health education and help with accessing financial services. It commits the city to a proactive role in achieving this – for example, it suggests that the city should subsidise training. Further, it suggests a sectoral or industry-by-industry approach to helping those working in the informal economy. This would entail comprehensive analyses of different sectors so the city can design and implement economically informed and targeted interventions. Although predating the CLEP, the policy suggests a very similar approach.

The policy drew from the lessons learned, particularly in the Warwick Junction Project, and was an attempt to institutionalise this learning. It was unanimously accepted by the Council, indicating a political commitment at the time to this shift in thinking. An Implementation Working Group (IWG) was established and initially made progress, including getting a lawyer based at the Legal Resources Centre (LRC), a human rights legal institute, to develop a new set of bylaws informed by the policy. These bylaws suggest that traders be encouraged to comply through a range of incentives, but the ultimate sanction would be the removal of the right to trade rather than fines and imprisonment – so securing the shift from criminal to administrative law.

The policy was, however, accepted at a time of institutional change. The 2001 local government elections formalised new city boundaries and ushered in a final round of local government restructuring. The policy had been developed for the former North and South Central Local Councils, only two of the six substructures that were amalgamated to form the eThekwini Municipality in 2001. Departments were merged and bureaucrats jockeyed for positions, and many left, including several active in the Informal Economy Policy development process. In 2002 the IWG merged with the Business Support unit. No progress was made in promulgating the new bylaws.

Pro-development politics: 2004–11

Policy in name but not in practice

The first signs of a significant shift in the city's approach appeared in mid-2004. On 14 June the Metro Police, without warning, removed traders' goods at various intersections throughout the central business district and the neighbouring middle-class suburbs. A Metro Police spokesperson was reported as saying, 'We have seized tons of their goods in our clean-up operation,' and adding, 'We won't let up until we have cleaned them all out' (*Daily News*, 15 June 2004). In November 2004 the Council approved the Public Realm Management Project (PRMP), designed to stop 'illegal, unlicensed street trading'. R3.7 million was assigned to the project over six months to be spent on employing 50 new Metro Police officers, who were armed but euphemistically named 'peace officers', a move justified by complaints from formal business, but the then head of the Business Support unit stated 'Illegal street traders are creating a terrible headache for shop owners and members of the public ... it is important to make sure that we also address the needs of the city's rates base by ensuring that the concerns of formal business are also being looked at' (*Daily News*, 9 March 2005). The Deputy Mayor, who also headed the economic development committee in the Council, confirmed this, 'While it is important to promote small businesses, we also need to safeguard the city's rates base by addressing the concerns of formal business' (*The Mercury*, 7 February 2005). At the time, the Council had issued a total of 872 permits, thus rendering 'illegal' the vast majority of street traders in the city. The 'peace officers' concentrated on removing traders throughout the central business districts of Durban and Pinetown, as well as the beachfront. Although the number of licensed traders has since gradually increased over time, street traders have been continually harassed.

Statements to the press indicate that the reason behind the Council's shift in approach was pressure from formal business. However, Marriott (2005), who interviewed a number of formal businesses operating in the inner city, concludes that 'while a minority of formal businesses reported occasional incidents of conflict between themselves and the traders, overall their relationship was described as cooperative' (ibid.: 13). Further, many street traders source their goods from formal businesses. Activists point to a correlation between these developments and the announcement in May 2004 that South Africa was to host the 2010 World Cup football event. The coordinator of the international alliance of street traders, StreetNet, who had been monitoring the situation closely, argues that the sudden clearing of the streets was partly because cities are competing against each other for football events. She points out that there was an unprecedented level of urgency with the PRMP. It was pushed through Council by the City Manager, bypassing all the usual committees (interview, Pat Horn, Coordinator, StreetNet, 13 September 2005).

These developments re-energised organising efforts among street traders. In December 2005, Siyagunda, an organisation of street barbers, who are predominately foreigners, was formed. Two further organisations mainly representing traders without permits – the Eye and CBD1 – became active again. In late November 2006, Siyagunda, the Eye and the Phoenix Plaza Association marched on the City Council and handed over a memorandum of demands to the Council's executive committee in an attempt to stop police harassment. In response to the success of the football World Cup bid, StreetNet launched a World Cities for All campaign, which in part aimed to coordinate these efforts. These developments also resulted in the two council officials who had led the Warwick Junction Project leaving the City Council in 2007 to establish a non-governmental organisation (NGO) Asiye eTafuleni (AeT). AeT means 'come to the table' in isiZulu and provides design, legal and other support to street vendors, market traders, barrow operators and waste pickers in Warwick Junction and elsewhere in the city.

Warwick Junction street and market traders under threat

To the surprise of many, in February 2009 the Durban/eThekwini Municipality announced its plans to build a shopping mall in Warwick Junction. The proposed site for the mall was the Early Morning Market (EMM), a fresh produce market in the centre of the Junction that was due to celebrate its centenary in 2010. The plans entailed a redesign of the whole district ensuring that the foot traffic, that was by this time estimated at 460,000 commuters a day, was directed past the formal rather than the informal traders, so threatening the viability of all street and market traders in the Junction.

Analysis suggests two critical issues behind this move. First, removing traders ahead of the 2010 football event was part of a plan to 'spruce up' the city. Second, the role of private property developers (who were, in this case, politically well connected) in driving the mall plans. In justifying the proposed development the City Manager noted that the area was in need of redevelopment, but also that, due to the football World Cup, the city was able to access national funds to reconfigure transport interchanges (*The Mercury*, 4 and 15 June 2009). Bromley (2000: 12) in his global review of street trading policy trends, notes that aggressive policing 'is particularly notable just before major public and tourist events, on the assumption that orderly streets improve the image of the city to visitors'. The City Council documents outlined that Phase 1 of the development would be completed before the commencement of the games. Although the City did not explicitly state that traders did not fit the image of Durban they wanted for the 2010 visitors, their actions certainly suggested this. This is part of a broader global trend where traders are often not considered to be part of a 'modern' city.

Warwick Mall (Pty) Limited was registered in 2006, demonstrating that the developers had been planning the project for some time. The company that established the consortium – Isolenu – specialises in township town-centre redevelopment. Given the high foot traffic in the Junction, for private retail this is a very lucrative and thus desirable site. The consortium approached the city with the mall proposal – in effect an unsolicited bid. In addition, the City notice of the intention to grant a 50-year lease of the EMM site to the Warwick Mall (Pty) Limited stated that there would be a one-off rental of R22.5 million (eThekwini

Municipality, 2009: 1). Later in the notice it stated that the municipality would contribute R24 million to the construction of a taxi bay on the roof of the proposed mall. The developer was thus not only accessing a very valuable piece of public land but was being paid by the city to do so. The City was thus facilitating private property developer interests. The consortium was well connected to local politicians. This echoes Harvey's (2008) analysis of the role of capital in urban development, which results in the displacement of the urban poor who happen to be located on centrally located land, a process of 'accumulation by dispossession'.

There was a groundswell of opposition to the proposal. Street trader organisations active in the area came together and were supported by the Congress of South African Trade Unions and the South African Communist Party in the KwaZulu-Natal province who publicly opposed the proposals. Civil society groups met weekly under StreetNet's World Class Cities for All campaign. Urban practitioners and academics also joined the campaign – writing letters to the press, arranging public debates and giving technical assistance.

Recourse to the courts

AeT worked alongside traders to launch a legal challenge securing the services of LRC. On behalf of a number of traders operating in the EMM, the LRC wrote to the City Manager on 2 July 2009, indicating their intention to oppose the demolition the market. The very next day – 3 July 2009 – the city issued a directive to the barrow operators stating that they were forthwith required to apply for permits to haul their barrows through the market. Barrow operators are central to the operation of the market. They ferry goods from suppliers to traders and from traders to customers. The LRC, at no cost to the barrow operators, launched an urgent application on their behalf. The Court ruled in favour of the barrow operators, striking down the city's attempts to curtail their livelihoods.

In the meantime, the LRC prepared a comprehensive set of court papers. A challenge to review the decision of the municipality to grant the lease for the EMM was filed in August 2009. The applicants were fresh produce sellers, barrow operators, chicken sellers and bovine head cooks. The review of the city's decision was based on administrative law and constitutional principles, ranging from aspects of procedural fairness in terms of the requirement to consult with a party likely to be adversely affected by an administrative decision, to arguments on the importance of preserving the site as a heritage landmark. In preparation for the application to review the city's decision, the LRC had to interrogate in great detail the Council's internal decision-making process. The decisions of the municipality were attacked on grounds that they were procedurally unfair, and made without genuine and meaningful engagement with traders and people who worked in and around the market, and with members of the public who would be affected by its closure. One of the grounds of attack was that the municipality appeared to short-circuit the normal procedures and time-fames associated with public tenders.

Since the LRC had outlined its intention to institute legal proceedings, the city had to halt its plans to evict EMM traders by the 31 July 2009 until the matter was resolved. This gave the traders an important respite, but also meant the development was unlikely to be completed before the World Cup events. The case was, in fact, never heard. By April 2011 the City Council finally rescinded its 2009 decision to lease the market land for the mall development, noting that there was 'little prospect of the legal challenges relating to the current proposal being resolved'. This was a major victory for the traders.

The 'Informal Economy Monitoring Study', 2012 (IEMS), undertaken by the action research network Women in Informal Employment: Globalizing and Organizing (WIEGO), gives critical insights into street traders' experiences of the ambivalence of local authorities to their activities (Mkhize *et al.*, 2013). This study included a survey of 150 traders and 15 in-depth focus-groups discussion with inner-city traders and traders operating in suburban areas. Access to essential basic and work-related infrastructure was identified as a critical problem – 56% of vendors surveyed did not have access to a toilet, while 21% did not have running water. The situation was worse in the city centre than in the outlying areas. Three-quarters of the vendors interviewed had no shelter while working so were exposed to rain, sun and wind, while nearly half of the vendors surveyed did not have access to storage. The focus group data showed that when vendors did have access to storage, their goods were often stolen or spoilt while in storage. When these findings are compared to the other four IEMS cities where vendors were also interviewed – Accra, Ahmedabad, Lima and Nakuru – Durban vendors have comparatively poor access to basic amenities. Access to infrastructure was identified as *the* major issue hindering vendor businesses, cited by 14 of the 15 focus groups (ibid.: 1).

Police harassment of vendors was reported to be pervasive. Three in every four vendors operating in the periphery, and one in every two operating in the centre, reported being harassed by the police. The second most serious negative force cited by the focus groups (after poor infrastructure) was the police. More than half (53%) of vendors surveyed in Durban identified confiscation of property as an important problem, more than any other city in the IEMS (the average was 32% across the five cities). When goods are confiscated, vendors reported that at best their goods are damaged – at worst, they are never returned at all. Vendors also noted that the fines charged are often onerous but vary from one officer to the next (ibid.: 2). Experiences are reflected in some of the quotes below.

> The police are harassing us, they take away our goods. We need to run away from them with our goods in order to be safe.
>
> Male vendor

> The police are a big problem, they are confiscating our goods and take our permits away.
>
> Female vendor

> We really suffer in their presence.
>
> Female vendor

Vending permits are clearly a critical issue. Half of the vendors interviewed in the survey cited not being able to obtain a business licence as a problem. However, as one male vendor stated, 'They confiscate our stock even if the permits are produced,' suggesting that having a permit does not necessarily protect traders from harassment. A number of vendors reported that police issue fines and/or confiscate their goods if the permit is valid but the permit holder is not there.

> Police are a problem. If they are at your table they ask for the permit. If someone else produces the permit on behalf of the owner, they say they need to see the person whose picture is on the permit. If the owner is not in the table, they issue a ticket for a fine of about R300 or R250 or R100, depending on the mood of that police officer at that time.

Confiscation of goods challenged in court

The Warwick Mall legal case has established a good working relationship between the LRC and AeT, who were increasingly seeing that litigation was one of the few routes of securing trader livelihoods. These two institutions were determined to challenge the pervasive confiscation of street trader goods but needed an individual trader willing to challenge the city.

John Makwicana was a trader leader who has been operating in Warwick Junction since 1992, earning a living for himself and eight dependents. He had a permit to trade and was authorised to employ an 'assistant trader'. Early morning on 6 August 2013, John set his goods out on display and then left to do some organisational work, leaving his assistant to tend his table. At mid-day, his assistant went to buy lunch from a nearby supermarket, and left John's trading permit with a nearby vendor. While the assistant was away, a police officer stopped at John's table. The nearby vendor called the assistant who rushed back, only to find the police officer packing 25 pairs of new sandals into a plastic bag.

The LRC and AeT decided to file a test case in the Durban High Court, challenging the power of the city to confiscate and impound the goods of traders, and seeking appropriate compensation for John. In the High Court, the LRC lawyer argued that the provisions in the eThekwini Municipality's street trading bylaws, which authorise confiscations, are in conflict with the constitutional right to equality, the right to choose one's trade or occupation, the right to property and the right of access to courts.

In February 2015, a judgment and an order from the High Court were made in John's favour. The presiding judge also ordered that the municipality compensate John with R775 (US$62), with interest, for the confiscated goods, and also compensate him for his legal fees. The judge declared the clause in the bylaw that allows for impoundment and confiscation of street trader goods 'unconstitutional, invalid and unlawful', and ruled that the city needed to redraft this section of the bylaws. This judgment represented a significant victory for Durban street vendors. The city is in the process of redrafting the bylaws, but has questioned national government's jurisdiction over street trading, indicating that further litigation is likely. In the meantime, Isolenu has secured the rights to build a shopping mall in another part of Warwick Junction.

In the recent era, where ambivalent attitudes to the informal economy have become entrenched, one of the few routes to securing livelihoods has been litigation. The Durban case suggests that having access to free, high-quality and responsive legal assistance for informal workers is key. This is particularly important in the face of powerful private sector interests. The role of intermediaries to find and support litigants was also important. CLEP rightfully draws attention to the supply side of the justice system.

Developments at national level

As the country-wide xenophobic attacks of 2008 and more recent xenophobic violence in Durban in early 2015 have shown, there is pervasive hostility to foreign nationals, with those working in informal retail being particularly vulnerable. In 2013, the National Department of Trade and Industry released a draft Businesses Licensing Bill to replace the 1991 Businesses Act. On the release of the draft, the minister was cited as saying the Bill was designed to 'deal with illegal traders and semi-illegal practices taking place in South Africa' (*Citizen*, 21 March 2013). The Bill specifies that any person involved in business activities – no matter how small – will be required to have a licence. The draft Bill is largely punitive, with harsh sanctions – no upper limits on the fines charged, someone found guilty of contravening the Act could be imprisoned for as long as ten years. The Bill states only foreigners who have a *business permit* will be granted licences. Business permits have to be applied for in the country of origin and are only granted if the person applying can guarantee that they have R2.5 million to invest in South Africa. Few if any foreigners currently operating in the South African informal economy will qualify. The draft bill was critiqued from many quarters and is currently under review. The widespread anti-foreign sentiment from local competitors, echoed by the ruling elite, is likely to reinforce the punitive approaches in the future.

Conclusion

This Durban case shows that the extent to which traders were and are incorporated into urban plans is a complex interplay between individual and collective street trader interests, the private sector and a heterogeneous state, containing both pro- and anti-integration elements. The case demonstrates that both national government and local law play a central role in these processes. As outlined in the introduction, CLEP provides a useful agenda for legal reform to secure street trader livelihoods.

The Warwick Junction Project of integration of street traders followed by Durban's Informal Economy Policy of 2001 addressed much of this reform agenda, and yet conditions for street traders were quickly reversed. Golub (2009: 101), in his critique of CLEP, argues that the Commission did not address how to best implement ambitious legal reforms, 'so that they have actual impact and do not simply exist on paper'. He points out that the Commission 'displays an inordinate faith in employing rational persuasion to grapple with a widespread problem it highlights – how to get self-interested politicians and other elites to forfeit their own advantage for the well-being of society'. Cousins (2009: 893) goes a step further, arguing that the approach ignores political economy realities. In the Durban case the interests of property developers facilitated by allies in the state quickly reversed gains secured by street traders and their organisations. This echo's Golub and Cousins' critique.

In unravelling the role of the private sector in shaping cities, Harvey sees urban social movements as the one source of hope, which he says seek to 'overcome isolation and reshape the city in a different image from that put forward by developers, who are backed by finance, corporate capital and an increasingly entrepreneurially minded local state apparatus' (2008: 33). The Durban case shows how litigation is one among many tools these urban movements need to employ. It also demonstrates the importance of groups like street traders having access to high-level legal assistance, to defend hard-won gains. Legal reform is a necessary but certainly not sufficient condition for inclusive planning for the working poor. A key priority in South Africa is ensuring the legal frameworks governing these activities are shifted from criminal to administrative law. In part informed by an anti-foreign sentiment but also a failure to recognise these activities as a part of the economy and modern cities, the current punitive approach is likely to continue.

Notes

1   Stats SA defines the informal sector as: employees working in establishments that employ less than five employees, who do not deduct income tax from their salaries/wages; and employers, own-account workers and persons helping unpaid in their household business who are not registered for either income tax or value-added tax (Statistics South Africa, 2015: xxii).
2   US$2.11 billion.
3   In 2000 the Warwick Junction Project was awarded the World Wildlife Fund South Africa and Nedbank's Green Trust Award in the category Urban Renewal. The Project Centre won a KwaZulu-Natal Institute of Architects Heritage Award for the renovation of the Centre – a listed building.

References

AFSUN (2016) The African Food Security Urban Network, http://www.afsun.org/, accessed September 2015.
Attwood, H. (2000) Assessment of markets in Durban's Central Councils, Research report for Durban's North and South Central Councils.
Bromley, R. (2000) Street vending and public policy: a global review, *International Journal of Sociology and Social Policy*, 20(1–2): 1–28.
Brown, A. and Rakodi, C. (2006) Enabling the street economy, in Brown, A. (ed.), *Contested Space: Street Trading, Public Space and Livelihoods in Developing Cities*, Rugby: ITDG.

Cousins, B. (2009) Capitalism obscured: the limits of law and rights-based approaches to poverty reduction and development, *The Journal of Peasant Studies*, 36(4): 893–908.

Crush, J. and Frayne, B. (2011) Supermarket expansion and the informal food economy in Southern African cities: implications for urban food security, *Journal of Southern African Studies*, 37(4): 781–807.

Data Research Africa (1998) Street traders survey of the Durban Metropolitan Area, Report produced for the Durban Metropolitan Council, Economic Development Department.

Devenish, A. and Skinner, C. (2006) Collective action for those in the informal economy: the case of the Self-Employed Women's Union, in Ballard, R., Habib, A. and Valodia, I. (eds), *Voices of Protest: Social Movements in Post-Apartheid South Africa*, Durban: University of KwaZulu-Natal Press.

Devey, R., Skinner, C. and Valodia, I. (2006) Definitions, data and the informal economy in South Africa: a critical analysis, in Padayachee, V. (ed.), *The Development Decade? Economic and Social Change in South Africa, 1994–2004*, Cape Town: HSRC (Human Sciences Research Council).

Dobson, R. and Skinner, C. (2009) *Working in Warwick: Including Street Traders in Urban Plans*, Durban: University of KwaZulu-Natal.

eThekwini Municipality (2009) Strategic Projects Unit and 2010 Programme, Notice of Intention to Grant the Lease of Immovable Property to Warwick Mall (Pty), Limited, www.durban.gov.za, accessed June 2010.

eThekwini Unicity (2001) eThekwini Informal Economy Policy, http://www.durban.gov.za/City_Services/BST_MU/Documents/Informal_Economy_Policy.pdf, accessed September 2015.

Fourie, F. and Kerr, A. (2015) Informal sector employment creation: what can the SESE Enterprise Survey tell us?, Paper presented at the Economics Society South Africa Conference, Cape Town, 3–4 September.

Golub, S. (2009) The Commission on Legal Empowerment of the Poor: one big step forward and a few steps back for development policy and practice, *Hague Journal on the Rule of Law*, 1(1): 101–16.

Grest. J. (2000) Urban citizenship and legitimate governance: the case of the Greater Warwick Avenue and Grey Street Urban Renewal Project, Durban, Paper presented to the South African Planning History Study Group Millennium Conference, 29–31 May, University of Natal, Durban.

Harvey, D. (2008) The right to the city, *New Left Review*, 53: 23–40.

Horn, P. (2004) Durban's Warwick Junction: a response. *Development Update*, 5(1): 209–14.

Hunter, N. and Skinner, C. (2003) Foreigners working on the streets of Durban: local government policy challenges, *Urban Forum*, 14(4): 301–19.

ILO (2002) Resolution Concerning Decent Work and the Informal Economy, ILO (International Labour Organization), adopted by the International Labour Conference at its 90th Session, June, http://www.ilo.org/global/docs/WCMS_080105/lang--en/index.htm, accessed September 2015.

KMT Cultural Enterprises (2003) An informal trader census in designated areas in the Durban Unicity, Draft report submitted to the Development and Planning Unit, eThekwini Municipality.

Ligthelm, A. (2006) Measuring the size of the informal economy in South Africa, 2004/5, Research Report no. 349. Pretoria, Bureau for Market Research.

Ligthelm, A. and Lamb, M. (2004), Size, structure and profile of the informal retail sector in South Africa, http://www.unisa.ac.za/contents/faculties/ems/docs/Press323.pdf, accessed May 2016.

Lund, F. and Skinner, C. (2004) Integrating the informal economy in urban planning and governance: a case study of the process of policy development in Durban, South Africa, *International Development Planning Review*, 26(4): 431–56

Marriott, A. (2005) Informal street traders and the public realm management project: perspectives of formal businesses in the Durban CBD, Report produced for the School of Development Studies, University of KwaZulu-Natal.

Mkhize, S., Dube, G. and Skinner, C. (2013) Informal economy monitoring study: street vendors in Durban South Africa, Inclusive Cities Research Report, Women in Informal Employment: Globalizing and Organizing, http://wiego.org/publications/city-report-IEMS-street-vendors-durban-south-africa, accessed September 2015.

Nesvag, S. (2000) Street trading from apartheid to post-apartheid: more birds in the cornfield?, *International Journal of Sociology and Social Policy*, 20(3/4): 34–63.

Reform Development Consulting (2010) A census of street vendors in eThekwini Municipality (Draft report), Report commissioned by StreetNet International.

RoS (1991) Businesses Act, RoS (Republic of South Africa), *Government Gazette*.

Rogerson, C. and Hart, D. (1989) The struggle for the streets: deregulation and hawking in South Africa's major urban areas, *Social Dynamics*, 15(1): 29–45.

Rogerson, C. (2004) Pro-poor local economic development in post-apartheid South Africa: the Johannesburg fashion district, *International Development Planning Review*, 26(4): 401–29.

Saunders, N. (2004) From ghetto to productive inclusion: Durban's Warwick Junction. *Development Update*, 5(1).

Skinner, C. (2008) The struggle for the streets: processes of exclusion and inclusion of street traders in Durban, South Africa, *Development Southern Africa*, 25(2): 227–42.

Statistics South Africa (2015) Quarterly Labour Force Survey, Quarter 2, Statistical release P0211, http://www.statssa.gov.za/publications/P0211/P02111stQuarter2015. pdf, accessed September 2015.

StreetNet (2011) Durban Street Traders, StreetNet International, http://wiego.org/publications/durban-street-traders, accessed September 2015.

UNDP (2008a) *Making the Law Work for Everyone: Volume 1, Report of the Commission on Legal Empowerment of the Poor*, New York: UNDP, http://www.unrol.org/files/Making_the_Law_Work_for_Everyone.pdf, accessed September 2015.

UNDP (2008b) *Making the Law Work for Everyone: Volume 2, Report of the Commission on Legal Empowerment of the Poor*, New York: UNDP, http://www.unrol.org/files/making_the_law_work_II.pdf, accessed September 2015.

UN-Habitat (2006) *Innovative Policies for the Urban Informal Economy*, Nairobi: United Nations Human Settlements Programme.

10 Resisting the revanchist city

The changing politics of street vending in Guangzhou

*Gengzhi Huang, Desheng Xue
and Zhigang Li*

Summary

Through a case study of transformations in Guangzhou's street-vending policy, this chapter explores how exclusionary practices of urban politics in China are undermined by those whom it seeks to exclude and a progressive political climate that questions the exclusionary framework. The exclusion of street vendors in Guangzhou was underpinned by the revanchist National Sanitary City (NSC) campaign, which proved difficult to operate due to the resistance of street vendors, who developed flexible, small-scale activism to maintain their livelihoods. Meanwhile, the national discourse of social harmony has driven local authorities to adopt an ambivalent policy approach, to mediate the tension between retaining an attractive city image and addressing the livelihoods of the urban poor in Chinese cities.

Introduction

The processes of socio-spatial exclusion of subordinated groups driven by punitive urban polices have been observed in contemporary cities around the world (Crossa, 2008; van Eijk, 2010). Spurred on by neoliberalism, globalisation and urban entrepreneurialism (De Verteuil, 2006; Macleod, 2002), the exclusionary policies have sought to erase urban spaces for some sectors of the population, such as street vendors, the homeless, beggars and sex workers. One particularly influential perspective in understanding such an issue has been the theory of urban revanchism first coined by Smith (1996). Revanchism represents the reclaiming of a city's prime spaces by dominant classes from 'deviant' social groups. Many researchers have taken up the issue of the applicability of revanchism to different urban contexts in the North and South (for example, Aalbers, 2010; Papayanis, 2000; Swanson, 2007; Schinkel and van den Berg, 2011; Uitermark and Duyvendak, 2008). In recent literature, however, there has emerged a discomfort with the revanchist assumption framing urban policies. Scholars have argued that the evidence of supportive approaches to address subordinated groups represents both an empirical and a theoretical counterweight to the revanchist thesis (DeVerteuil, 2006; Johnsen and Fitzpatrick, 2010;

Laurenson and Collins, 2007). DeVerteuil *et al.* (2009a) thus suggest that the nature of urban political responses to the subordinated should be understood as multifaceted and ambivalent rather than only punitive/revanchist. The ambivalence mixing punitive and supportive strategies has been termed as the 'post-revanchist' approach (Murphy, 2009; Mackie *et al.*, 2014).

This chapter aims to understand the development of punitive urban policies by focusing on the roles of the counter forces to exclusionary practices of revanchism. As Smith (1998: 17–18) suggests: 'Top-down revanchism will not go unchallenged ... it is equally important to retain a sense of how alternatives might emerge, where the sparks of change might come from.' In the literature on the understanding of revanchism practices in the global South, the street-vending problem has been a particularly apt litmus test (Bromley and Mackie, 2009; Crossa, 2008; Popke and Ballard, 2004; Swanson, 2007).

In Guangzhou, the recent change in street-vending policy relates to a critique of revanchism that suggests ambivalent forms of urban policies for the subordinated. Since 1990, the management of street vendors has been directed by the exclusionary approach driven by the 1990s National Sanitary City (NSC) campaign as a revanchist project that seeks to eradicate the undesirable from public space. Recently, the exclusion-oriented approach has been mixed with the provision of inclusionary vending places, which suggests an alternative future for street vendors. The key question addressed by this chapter is: what drives the transformation of exclusionary street-vending politics to an ambivalent or post-revanchist approach? In a sense, it is an attempt to take up Smith's suggestion by conducting an empirical examination of the ways in which the exclusionary urban policies are undermined and reshaped with alternatives in Chinese cities. Specifically, we will examine how street vendors, the group facing official removal, resist and challenge the exclusionary policy, and how this challenged policy is tempered and reshaped by the emerging discourse of social harmony at national level. By examining the characteristics of the inclusionary approach, we also reveal the nature of ambivalent street-vending politics in Guangzhou.

The chapter starts with a review of the literature on urban revanchism, with an emphasis on understanding the counter forces that it faces. In the subsequent

Figure 10.1 The ambivalent context of street vending in Guangzhou.

sections, we first examine in what sense the exclusion of street vendors in Guangzhou is understood as a representation of revanchism, and then explain how the resistance of street vendors, and the 'harmonious' discourse, both challenge and reshape exclusionary street-vending politics. Conclusions follow in the final section.

Understanding urban revanchism

Urban revanchism was fuelled by the economic recession and a discourse of 'spiralling urban decay' in the 1990s in New York City, which had dismantled its liberal urban policy since the 1980s (Smith, 1998). Revanchism essentially represents the city's intent to eradicate undesirable populations from prime urban spaces in order to create positive images, thereby attracting highly mobile capital in an era of intense interurban competition (DeVerteuil *et al.* 2009a; Johnsen and Fitzpatrick, 2010). Revanchism operations are practised through actions concerned with removal and zero tolerance (Smith, 1998), the punitive or coercive purification of public space (DeVerteuil, 2006) and the employment of illegalisation, discursive stigmatisation and physical manipulation of spaces and their surveillance (Johnsen and Fitzpatrick, 2010). Embodying 'the ugly cultural politics of neoliberal globalisation' (Smith, 1998: 10), revanchism performs a principle of the neoliberal city, namely that the government will repress some segments of the population if there is a conflict between creating a good business image and their personal wellbeing (Harvey, 2007). However, rather than existing in a homogeneous form (Smith, 2001), adapted versions of revanchism have been explored through applied research in different urban contexts (van Eijk, 2010). They assume 'actually existing' revanchism (Macleod, 2002) in the context of the emergence of 'actually existing neoliberalism' (Brenner and Theodore, 2002). Attempting to detect different manifestations, Atkinson (2003:1833) identifies competing strands in the understanding of revanchism: the economic objectives of attracting capital investment, a prophetic and dystopian image of urban decline, a set of programmes to secure public space and a mode of governance to dictate recognised uses for urban spaces. It is therefore possible to speak of a certain 'degree of revanchism', given that policies more or less fit these strands (van Eijk, 2010).

Downplayed in the literature, however, are the ways in which those who face removal resist, challenge and even subvert the exclusionary practices of revanchism (for example, Crossa, 2008). In the light of Lefebvre's notion of social space, revanchism could be understood as the governmental conceived and regulated *representation of space*, which is in contradiction with *representational space* actually used by various groups and individuals (McCann, 1999). The contradiction represents the class struggle between the dominant and subordinate, exposes counter voices and opens up possibilities for alternatives. Revanchism is thus inherently *unstable*. Consequently, Macleod suggests the need to: 'to explore the spatializing practices and "counterspaces" of resistance and transgression that can sometimes unshackle the padlocks of "purified" urban sites and thereby challenge their official, growth-machine-dominated representations of space' (Macleod, 2002: 618).

Nuanced forms of resistance to structural constraints by the subordinated – other than political and social movements – have been explored (Bayat, 2000; Kerkvliet, 2009). Scholars have distinguished the political and collective actions mobilised by street-vending organisations and more flexible, individual and small-scale activism in different contexts (Bayat, 2000; Cross, 1998a). Bayat (1996) identifies a mixed form of resistance as 'quiet encroachment' that involves everyday silent, prolonged and atomised actions, and the episodic collective protest without structured organisation. In addition, Crossa (2008) highlights street vendors' *torear* resistance through mobilising to resist the entrepreneurial city. Flexible activism is also celebrated by DeVerteuil *et al.* (2009b), who explore a range of homeless resistance behaviours along a continuum from exit, and adaptation, to persistence or voice (social protest).

Emphasising the roles of the resistance, this chapter first discloses the structural roots of street vendors' persistence in modern China to understand the necessity for their resistance – a question rarely covered in discussions on the counter responses to revanchism. It then differentiates vendors' resistance strategies for escaping, bribing and confronting the powerful, and subsequently explains their different impacts on the policy. Nevertheless, the resistance here is generally understood as a form of resilience, a range of actions opposing the exclusionary policy in order to survive rather than to effect change. This builds on Katz's significant distinction between different countering responses. Katz (2004) contrasts forms of *resistance* that involve subversive actions and emancipatory consequences, and forms of *reworking* that alter oppressive and unequal circumstances, with *resilience* that enables people to survive without really changing exploitive and oppressive conditions. In effect, resilience accords with Cresswell's 'transgression' that diverts and manipulates rather than deconstructs the power of established boundaries and spaces (Cresswell, 1996). This definition of resistance allows us to understand, at least in this study, why the counterforce from below is insufficient to effect policy change.

In examining the unfolding of revanchism, scholars have argued that urban governmental responses to subordinated groups are not always framed by the forces of neoliberalism. Murphy (2009) argues, for instance, that a progressive political climate that frequently undermines the neoliberal imperatives renders it unacceptable to simply remove the poor. This gives rise to an ambivalent new benevolence in new urban polices. Likewise, in their critique of the punitive city, DeVerteuil *et al.* (2009a) contend that the fact that the state cannot be reduced to the handmaiden of capital leads to a possibility for developing supportive policies for the poor. In China, though neoliberal shifts have penetrated urban development since the introduction of market-oriented reforms (Liew, 2005), it is argued that neoliberalisation processes are inconsistent (He and Wu, 2009). Harvey (2005) further asserts that the Chinese state has to depart from neoliberal orthodoxy and to act like a Keynesian state if it is to achieve social and political stability. Nonini (2008) similarly argues that the Chinese state is trying to keep a dialectical balance between capitalist accumulation and socialist values. Indeed, the Chinese state's concern with this balance is manifested in its recent decision about 'building a harmonious society', which has required local governments to resolve social conflict in ways that benefit social stability and take people's livelihoods seriously. Our interest is how the national discourse on social harmony mediates and reshapes exclusionary street-vending politics at local level.

Methods

The choice of Guangzhou arises from its political position on street vending in China. The management of a huge number of street vendors (about 0.3 million) has been a focus of urban policies and has received considerable national attention, and Guangzhou is also a pioneer city in China that has transformed its street-vending policy. This transformation was not observed in other cities. We do not claim that the city's street-vending problem mirrors the national profile, given considerable regional differences among Chinese cities, but rather that it offers a relevant and significant case for understanding the possibilities of, and driving forces for, the change in exclusionary urban policies. As our interest is in using an in-depth case study to examine how the exclusionary practices of revanchism are challenged and transformed, we do not aim to theorise generally about urban revanchism in China. Rather, as part of this research, we detected elements of revanchism from evidence of the exclusion of street vendors in Guangzhou. We approached this task by examining the relationships between the objectives of urban politics, the NSC campaign and the exclusion of street vendors. Specifically, we explore the questions concerning what drives the NSC campaign and how exclusionary practices are mobilised.

This research involves a long-term observation of Guangzhou's street-vending issues. We triangulated a wide range of materials drawn from interviews, participation observation and archival research. In 2007 and 2012, we spoke with urban management officers at different levels of administration, including municipal officers involved in the policy decision-making process and the district and sub-district officers who implement the policy, in order to understand the government's motives in making and adjusting the exclusionary policy, and to what extent vendors' resistance had affected the policy. From October 2011–February 2012, we conducted semi-structured interviews with 200 street vendors in 20 main vending locations in Guangzhou. We selected the samples based on the gender of the vendor and types of goods sold in order to analyse the diversity within this social group. Asking questions about what drives them to engage in street vending and how they experience the old and new policies, we sought to understand the need for their resistance, the ways in which they challenged exclusionary practices and the characteristics of the inclusionary strategy. This research was further supplemented by archival research and analysis of newspaper articles to examine the development of the NSC campaign, the effects of the 'harmonious' discourse and the formation of the ambivalent approach.

National Sanitary City campaign and the exclusion of street vendors

In socialist China (1949–77), street vendors were viewed as the 'tail of capitalist economy' and forced into cooperative groups and cooperative stores as part of the socialist economy. Since the reform and opening-up in 1978, they have proliferated in cities due to rapid urbanisation and deregulation of the private sector. Meanwhile, the responsibility for street vendors has shifted downwards to local governments in the general process of power decentralisation and market-oriented reforms (He and Wu, 2009). Management of street vendors has become an urban issue that is not considered in national policy. In the 1980s, street vending was generally tolerated by city authorities because of its significance in alleviating serious poverty and supporting the undeveloped urban retailing system (Bannan, 1992). Since the 1990s, however, street vendors have been faced with total exclusion as a result of the NSC campaign, launched in many Chinese cities. The campaign was initiated by the National Patriotic Health Campaign Committee in 1989 to encourage local governments to improve deteriorating urban environments. Cities are entitled to become an NSC if they meet the given criteria. Although it is not a mandatory policy, there have been an increasing number of cities engaging in the campaign because of its economic and political position in urban politics.

Drivers for the NSC campaign

As the development pressure confronted by the central state is transferred to the local state in the process of market transition (Wu, 2002), economic and urban growth becomes the overriding goal of urban politics in post-reform China. Meanwhile, local authorities are now capable of using various strategies to attract investment and to regulate local development due to power decentralisation. One of these strategies has been the competition for local and global capital. Through public investment in infrastructure, designation of development zones, preferential treatment to investors and land-leasing instruments, municipalities have engaged passionately in the enhancement of city images to create favourable investment climates (Wu, 2003). The NSC campaign[1] represents part of the image-enhancement strategy in Guangzhou, and other Chinese cities.

In Guangzhou, the image-enhancement strategy was adopted as a political response to its slowing development in the 1980s. From 1980–9, the proportion of Guangzhou's GDP in the total of Guangdong Province decreased from 23.1% to 20.8%; during the same period, the comparative annual growth rate of GDP in Guangzhou reduced from 15.4% to 4.8%. This does not suggest, given its positive economic growth, that Guangzhou experienced urban decline, rather that its economic competitiveness was falling compared with other cities in the province, especially in the Pearl River Delta (PRD) region. As urban planners asserted, Guangzhou might have been losing its central status in the region because it was faced with internal growth constraints and intense external competitiveness

(Wang *et al.*, 1993). As a response, the municipal government put forward the strategy of environmental improvement in order to construct a modern city with images of a beautiful environment, good order, civilised citizens, a booming economy and liveability (Lin *et al.*, 2005). To assist in promoting this strategy, the government established a guideline in 1998, called 'three times change', meaning 'in one year a small change, in three years a moderate change, in 2010 a big change'. By achieving these changes, Mayor Lin declared: 'The central status of Guangzhou in PRD is unshakable and Guangzhou will become the most liveable city with the capability of accumulating wealth' (Wang, 2000).

The NSC campaign was launched in 1990 in this context and has been promoted strongly since the late 1990s. The city's new Mayor Zhang has repeatedly emphasised that the campaign is necessary in achieving a 'big change' in city image, and to accelerate the building of Guangzhou as a modern international metropolis. The campaign is positioned as a key device to enhance the city image, improve the investment climate and thereby accelerate economic growth (GMG, 2003: 18). In the campaign, the government commits to a series of actions to beautify city landscapes (for example, keep public space clean and tidy), to protect the environment (for example, control pollution below a certain level), to guarantee food safety, to eradicate the pests (rats, flies, cockroaches, mosquitoes), to prevent infectious disease and to promote good hygiene. It thus aims to reshape the urban environment that had long been characterised by images of dirt (*zang*), disorder (*luan*) and badness (*cha*), and to create a sanitary, beautified and orderly city with the NSC title to benefit city branding. However, the Guangzhou municipal government have long worried that other PRD cities, such as Shenzhen (one of Guangzhou's main competitors), have already secured the NSC title. These cities were found to attract increased inward investment after entering the NSC club (ibid.: 11), making it more urgent for Guangzhou to promote the campaign in order to avoid the loss of economic competitiveness.

The NSC campaign is therefore driven by economic motives to accumulate capital. Nevertheless, unlike Western cities attempting to reverse economic recession by spatial purification (for example, Aalbers 2010; Atkinson 2003; Smith 1998), the campaign is driven by the city's intent to accelerate economic growth. Moreover, in accord with the argument that the exclusionary policies are not only driven by economic factors (Uitermark and Duyvendak, 2008; van Eijk, 2010), the campaign is also fuelled by a political motive held by local politicians to enhance political performance. In China, a city's mayor is not elected by the citizens, but appointed directly by higher-level government. Thus local politicians are much more accountable to the higher-level government than the people. Hence, urban politics is largely directed by local politicians' pursuit of political performance. As the success in the campaign is counted as a political achievement and may even result in a direct promotion, local politicians are enthusiastic about pushing it forwards, often with stark disregard for people's interests. As a result, some problems hindering the campaign (for example, street vendors) are simply removed if they cannot be solved within a politician's five-year term of office.

The exclusion of street vendors

Thus, we consider the NSC campaign to be a revanchist project, which seeks to eradicate undesirable groups from urban public spaces to satisfy the economic and political intents of the local ruling class. In the sense that street vendors are a group explicitly targeted, their exclusion is viewed as a representation of revanchism. Like the revanchist discourse in other contexts (for example, Smith, 1998; Swanson, 2007), street vendors are identified as a *sign* of dirt, disorder and low quality, seen to litter urban landscapes, produce unsanitary foods, obstruct city traffic and disturb the market order. They are thus incompatible with the image of the NSC. Although the campaign has affected the lives of other populations (for example, the homeless, beggars), it is the street vendors that are blamed as 'a lion in the way'. They are viewed as a major cause for the failure of the campaign, given their large number and visible spatial impact. Therefore, based on the regulation of city landscapes, statements concerning 'no street vending', 'no disorderly hawking' and 'no hawkers in public space' are repeated in the plans and working programmes of the campaign.[2] Attempting to strengthen the control, the government added a penalty clause[3] to the street-vending clauses in the new city landscape regulations in 2007, creating an 'annihilation of space' bylaw (Mitchell, 1997). Furthermore, street vendors were perceived as 'invaders' that do not belong to Guangzhou. A municipal urban manager declared: 'Today if we do not restrict them, but provide them with 10,000 jobs, then tomorrow there will be 100,000 vendors coming over [to] … Guangzhou, if undefended … even the fool will come to Guangzhou' (*Yangcheng Evening News*, 2009).

This hostile voice has its seeds in China's household registration (*hukou*) system that divides the urban population into local residents, who have the right to welfare, and migrants, who do not. The former often fear that the latter will grab the limited public resources, such as jobs or education. Given that most street vendors are migrants, it is understandable that some citizens embrace the sentiment held by the city manager: 'They evade tax and occupy limited public resources without contribution to the city. Why do we welcome them? They should earn their living in legal ways; otherwise, the injured are Guangzhouese.'[4] Therefore, while ethnic and racial factors fuel the hostile stigmatisation of subordinated groups in the revanchist practices in Western and Latin American contexts (Swanson, 2007; Uitermark and Duyvendak, 2008; van Eijk, 2010), in China, the *hukou* system functions as a binary population-division that gives implicit support to the perpetual exclusion of street vendors perceived as 'outsiders'.

The NSC campaign declared a 'spatial war' against street vendors. The Urban Management Composite Law Enforcement Bureau (UM or *chengguan*) was established in 1997 to enhance the control of public space and to take direct charge of the campaign. In particular, the UM is given responsibility by the local government for repressing street vendors with zero tolerance from the departments of police, taxation, business and sanitation. The UM officers carry out daily policing of public spaces to restrict street vendors, using electronic monitoring and patrol techniques. Moreover, periodic repressions with the slogans of 'Hundred Days' Action' or 'Spring Action' are mobilised to eradicate street vendors. Hundreds of UM officers, sometimes with the participation of the police, taxation, business and environment departments, suddenly attack selected areas, hitting street vendors by forcefully confiscating their goods. These actions are often mobilised just before an NSC appraisal or prior to mega-events such as the Asian Games. However, these exclusionary practices are challenged and negotiated by street vendors, who develop various survival strategies.

Street vendors' resistance: forms and effectiveness

Street vendors' resistance as a necessary act in transforming China

Street vendors' resistance is driven by the force of necessity – the necessity to survive and improve their lives. This necessity arises from the effects of the particular socio-economic transformation characterising post-reform China. First, street vending is the only means of survival both for labourers who have been released due to the reform of state-owned enterprises since the mid-1990s, and for marginal groups such as the disabled, homeless and ethnic minorities (for example, Miao or Uygur ethnic groups) that are excluded or discriminated against in an increasingly segmented urban labour market. For instance, one laid-off worker complained that 'I am old now and have no way out. Doing this is just for surviving, a way of "begging for food", like stealing, robbing or begging' (interview, 2011).

Second, street vending is also an optimal way of supplementing the income for several groups of people: 1) peasants confronted with rural poverty resulting from the state's development bias towards urbanisation, and who move to urban areas in the slack farming seasons to work because, as many said, 'farming makes no money' or 'farming is only enough for eating'; 2) rural migrant workers, forced by extreme working conditions (long working hours, low wages, insecure workplaces and inhumane factory regulations) in low-cost, low-end and labour-intensive industries, who quit their jobs to seek better livelihoods; and 3) urban workers, who are hardly able to survive and support their families in conditions of increased income disparity and creeping inflation in globalising cities, such as Guangzhou.

Despite the heterogeneity of these groups, street vending is usually a last resort in securing better living conditions, and a common sentiment expressed was that 'I don't know what to do if I don't do vending.' Confronted with the prospect of removal, street vendors have no alternative but to take up resistance as a *natural* act to protect a livelihood – selling in the streets – that enables them to adapt to transforming social processes.

Forms of resistance

With the absence of organisational power, however, street vendors cannot oppose the policy by negotiating with the government through formal means. They are banned by the government from joining any formal association, such as the China Private-Owned Business Association, because they are unlicensed. Nor can they establish their own associations due to restrictions on civil society organisations in China. Consequently, they demonstrate a flexible, atomised and small-scale activism to survive exclusionary practices.

Everyday non-violence: escape and bribe the powerful

The most prevalent strategy practised by street vendors is everyday non-violent resistance. This is characterised by actions of mobility and temporary compromise to escape the dominating powers, like the action of *torear* in Mexico City (Crossa, 2008) and de Certeau's 'spatial tactic' (Gardiner, 2000). Vendors keep vigilant and are ready to flee and hide in nearby safe areas (for example, alleys, backstreet, hospitals) to escape the UM's routine inspection. Without physical clashes, they pretend to leave, but in practice either move to the next street to sell or wait and return to their original places after the inspectors leave. One of them said: 'Retreat when the enemy attacks (*Di jin wo tui*). Everybody here knows this. *Chengguan* may come at any time; you must keep watch constantly, and be ready to run quickly … Don't fight against them, otherwise you will suffer more losses' (interview, 2011).

To reduce the risk of exposure to the inspectors, some vendors choose to operate their business when the officers finish work between 12 noon and 2pm or after 6pm. In order to adapt the strategies of retreat and return, they make and remake their equipment. Some attach their products to their body and sell by using a hand-held basket, while others place their products on a bicycle, mini-trailer, blanket or a wooden shelf that can easily be opened and folded. This creative practice enables them to switch constantly between selling and moving, making their behaviour difficult to define as 'occupying public space'. For instance, in prime spaces such as commercial areas and railway stations, the vendors stop repeatedly for a moment to sell, but meekly move on when running into inspectors.

Moreover, street vendors in a given locality spontaneously warn each other about the inspectors by shouting out '*zougui*', literally meaning 'escape the ghost'. This warning system represents a 'passive network' whereby atomised individuals tacitly recognise their common identity by mediating of space (Bayat, 1996). By using this network they get to know each other and talk about safe or risky places and the UM activities. This network is strengthened by the relationship of *laoxiang* (fellow townsmen) which is pervasive in Chinese society, and means that street vendors from the same region (town, city or province) recognise each other and engage in practices of mutual self-help. For example, some sell goods and some are sent to watch the streets and inform others about UM activities by using walkie-talkies or mobile phones. If one is caught by the officers, the others will attempt to disrupt the enforcement by gathering together to create a disturbance.

Some vendors also attempt to penetrate the powers by bribing the UM assistant officers, who are responsible for daily surveillance in a given locality.

These officers are easier to bribe because they are not trained as well as senior officers, and they welcome the opportunity to supplement their low wage. Street vendors bribe them on a daily basis with cash (for example, 5–10 yuan per day) or cigarettes to obtain permission to sell. Sometimes these assistant officers divulge information about UM activities in advance. Nevertheless, in many cases, street vendors are also asked by these officers to pay for their tolerance. Though such payment is unauthorised and illegal, many vendors interviewed said that they prefer to pay because the payment can be offset by the gain from selling without harassment.

Episodic violence: confront the powerful

Violent confrontation is less prevalent, but an important resistance practised by street vendors defending their gains from the UM's violent and coercive actions. Individual vendors, particularly those who frequently have their goods and equipment confiscated, use weapons such as knives, steel pipes or cooking oil to beat back the UM officers attempting to confiscate their goods. Moreover, the vendors may harm themselves, use rope to bind their hands to the law-enforcement vehicle and stand in the way or on the top of the vehicle in order to stop the officers leaving and to claim back the confiscated goods. Though these direct confrontations result in individual vendors being penalised, injured, imprisoned and even killed, they may choose this action as the only way to protect their livelihood. Many of them agree with such action:

> *Chengguan* are so brutal; like a bandit. I don't remember how many times I have fought against them. I neither steal nor rob, but why do they take away my goods. This is my living. I have two children to support for their education. I can't stand it!
>
> Interview, 2012

Small-scale collective activism is also mobilised by vendors united through the *laoxiang* relationship. A couple of vendors in the survey used an existing *laoxiang* relationship to confront repressive actions by the government, such as bursting into the UM office to reclaim goods and making a premeditated assault on UM officers for revenge. Collective actions may also be launched to protect existing gains – for example, in 2008 the government authorised a property management company to take over the management of Baohua Road within the historic centre, where street-vending activities have long existed. The vendors in this area are well networked as most of them come from Chaoshan and Dianbai regions. The company declared that street vendors should be cleaned up, except for those who pay for management fees of 500-1500 yuan per month. However, this plan caused a collective protest from hundreds of vendors, which led to a violent conflict between vendors and the company's security guards. Consequently, the government was forced to abandon this measure as a result of the protests.

Effectiveness of resistance

Although the resistance of street vendors is the result of their autonomous initiatives to maintain a livelihood rather than an attempt to overturn the established rule, it presents unexpected challenges to the exclusionary practices. First, the non-violent resistance weakens the effects of the UM's routine surveillance. The authority tried to extend the policing hours to respond to the spatio-temporal flexibility of street vendors, but this was ineffective because a resurgence of vendors always followed the end of policing periods. A long-term measure was the establishing of a team of assistant UM officers to reinforce surveillance in the streets, but these officers tended to extort bribes or to be bribed by street vendors. Through bribery, vendors build a patron–client relationship with the surveillance officers, thereby carving out a temporary 'safe space' in the official control system. This relationship presents the problem of 'weak state integration' (Cross, 1998b), which cripples the policy in practice. As a result, ironically, it was often found that the UM officers made an unwritten compromise with street vendors by warning them in advance not to appear in specific locations during the course of the NSC appraisal. This demonstrates that a partial and temporary alteration has been made to the operation of the exclusionary policy.

Violent confrontation has forced the authorities to acknowledge that suppression is a battle in which both sides lose. It was said that the five years from 2005–9 witnessed 2,626 cases of violent resistance by street vendors, with 1,679 UM officers hurt. More importantly, the confrontation resulted in many incidents of street conflict, which exposed the ugliness of the revanchist politics. These incidents have attracted continuous attention from the news media and have aroused wide public concern. With the increasing exposure of violent incidents, particularly those involving collective protest and the death of vendors, the UM has been censored for abuse of power and labelled as 'cold-blooded' and 'ruthless' and named a 'bandit' in public opinion. Public attitudes towards street vendors are often sympathetic as they are considered to be 'the disadvantaged' and 'the poor', fighting for their 'survival'.

Furthermore, incidents of conflict have provoked local reflection on urban development and a desire to see the street vendors treated with kindness. 'Why should street vendors pay a sacrifice for the National Sanitary City?', 'Does the city's right to be clean outweigh the street vendor's right to live?' and 'Is city image more important than people's livelihood?' are among the public questions that challenge the crackdown policy. As a response to public pressure, the authorities have had to advise enforcement officers to retreat when confronted with violent resistance, both to reduce the conflict and the officers' risk of injury. Thus, although the resistance of street vendors renders the exclusionary strategies difficult to implement and produces a counter discourse against the legality of the policy, its limit lies in the fact that the policy has proved unable to subvert established circumstances. As explained below, the emerging discourse of social harmony plays a crucial role in driving the policy transformation.

Harmonious society and ambivalent street-vending politics

Harmonious discourse and its influences on street-vending politics

The growth-first strategy pursued by the Chinese state has caused various social problems, including uneven regional development, unequal distribution of public resources, a widening income gap, rural poverty and urban unemployment. These problems have given rise to serious social contradictions which the Communist Party of China (CPC) thought might shake social stability and block future reform (Chen and He, 2008: 408). The CPC realised that, as China's GDP per capita increased from US$1,000–US$3,000 (achieved in 2004), China had reached a crucial development stage with the coexistence of golden development opportunities and rising social contradictions. The latter must be taken seriously to secure the former. Therefore, the CPC made a major policy decision seeking to 'build a harmonious socialist society' in 2006. The core objective of a harmonious society is to maintain social consensus and stability. As Chairman Hu Jintao claimed, the party and people should 'take more initiatives to face and resolve contradictions, increase harmony as much as possible, decrease disharmony as much as possible' (ibid.: 410). This ideology of harmony actually represents the intention of the state party to alleviate the inherent tensions and controversies of neoliberalism, and the key approach to alleviating social contradictions and achieving social harmony is 'to address the most direct and the most realistic interest problems that the people are most concerned about' (ibid.: 414). Hence, social justice and people's livelihoods (*minsheng*) should be taken as seriously as capital accumulation. Hu's 'social harmony' soon became one of the focuses of urban politics. For instance, Guangzhou Municipal Party Secretary Zhu highlighted the importance of social harmony in a major 2006 report of CPC Guangzhou's Ninth Congress entitled *Building Harmonious Guangzhou*.

In urban politics, the discourse of social harmony has required local governments both to resolve social conflict in ways that benefit social stability rather than intensify conflict, and to pay attention to people's livelihoods in addition to the goal of rapid economic and urban growth. Accordingly, the exclusionary street-vending politics is questioned on two fronts: first, that the approach triggers street conflict and increases elements of social instability due to the practice of violence; and, second, that it completely denies poor people a livelihood given the zero tolerance of street vendors. Nevertheless, as a result of the devolution of street-vending regulation, there are various local responses to the demand for social harmony. In Guangzhou, the process by which exclusionary politics is mediated and reshaped by the new political climate is embedded in the progressive responses of local authorities to the issues of people's livelihoods, mitigation of social resistance and harmonious urban management.

Two months after the 2006 decision to build a harmonious society, in an open meeting sponsored by a local media institution, the *New Express*, on the topic of 'unlicensed vendors and urban management', the UM authority insisted that the exclusionary approach would continue for securing order in public space, despite its limited effects. This standpoint was opposed by a provincial Chinese People's Political Consultative Conference (CPPCC) member, Meng. He contended that the government should be responsible for the livelihoods of street vendors as disadvantaged individuals, and should resolve problems such as food hygiene and spatial disorder rather than simply erase the activity itself. This argument questioned the revanchist ideology that seeks to remove rather than tackle social problems in order to 'clean up the city' (Atkinson, 2003: 1831). As Meng critiqued, this ideology 'only cures the symptom but not the diseases' (Zhang, 2006). In line with the wishes of street vendors who supported Meng's view, the solution should lie in the opportunity to use spaces as a basic resource for vendors' livings. However, the authority responded that practically it was beyond the government's capacity to support thousands of street vendors who are mostly migrants. Nevertheless, the idea of support had started to emerge in similar meetings with a rising concern for people's livelihoods in urban politics.

Meanwhile, the street vendors' resistance, particularly violent confrontations provoked by punitive practices, had been considered increasingly unacceptable in harmonious discourse. In 2008, the Standing Committee of Guangzhou National People's Congress (NPC) launched a consultative meeting in an attempt to mediate the tension between street vendors and urban management. The UM's violence, denounced by the vendors at the meeting, was taken seriously by the authorities, who promised to decrease violent actions through improved training of enforcement officers. However, the more fundamental concern is that of extinguishing the possibilities for social resistance. The provision of permitted vending spaces was again proposed as a way of changing the contradictory relationship between urban managers and street vendors, and also as a way of improving people's livelihoods. The authorities did not oppose the idea as they had in 2006, largely because of the unsustainable nature of the exclusionary framework in the new political climate. As a municipal urban manager said, 'complete suppression does not really work in practice because of constant violent resistance. You can't help it. If you coercively suppress, they may desperately fight against you. It just leads to violent conflict, going against social harmony' (interview, 2012).

In 2009, partly responding to the continual contention over street-vending politics, Secretary Zhu made an important written instruction requiring the UM to 'explore a new path of constructing a lawful, people first, civilised and harmonious urban management' (Luo and Huang, 2011). This made it imperative for the UM to seek alternative street-vending policies that fitted the new direction of urban management. In addition, with the purpose of guaranteeing social stability, the central state required local states to pay attention to urban unemployment resulting from the 2008 global financial crisis. This requirement has also urged a new approach to street vending as a way of employment. However, while the idea of inclusion was on the agenda, the problem was to what extent street vendors should be included in urban spaces given that (as the authorities worried) full toleration would contaminate the city's image, disturb urban order and undermine the NSC goal (Guangzhou UM, Liwan District Branch, 2010). This concern revealed the contradiction between elements of social harmony and principles of urban development. A solution to the dilemma was then proposed in an official report, namely to 'treat stubborn disease by setting permitted space and promote social harmony by mixing inclusion and exclusion' (Guangzhou Urban Management Committee, 2009). The report stated:

> From the perspective of Mao's Contradiction Theory, the contradiction between urban managers and street vendors is the one among the people. We should adopt the measures of education, guidance and accommodation, rather than 'you chase I run' like 'cat and mouse'. Therefore, in order to solve the contradiction between the notions of 'purify the city image' and 'fill the belly', we advocate an approach with a mix of inclusion and exclusion.

This ambivalent approach functions by including street vendors into some areas of the city, but excluding them from others. It reflects the needs of local government to both mitigate conflict and take people's livelihoods seriously and to secure an attractive city image to attract capital investment. In the words of the UM authority, 'the municipal government aims to acquire a balance between *mianzi* (city face) and *lizi* (people's livelihoods)' (interview, 2012). Hence, the ambivalence addresses the NSC goal by retaining the exclusionary strategies that signify that Guangzhou is not a paradise for street vendors or 'foolish' migrants. Meanwhile, the extent of inclusion can be controlled by defining who can be included, how many and where inclusion can occur. This thereby addresses the government's concern about its inability to solve everybody's livelihoods.

Characterising the ambivalent politics

According to the 'Forbidden area for street vendors' notice issued by the UM authority in 2010, street vendors will continue to be suppressed without exception in the area shown in Figure 10.2. It consists of 250 zones in the city, including 145 main roads, 56 key places and 49 sites around the Asian Games stadiums. These zones are defined as 'key areas' in representing the city image, such as the administration area (the historic city centre), city parks, central business district (CBD), the exhibition centre and train stations. In struggling to achieve harmonious urban management, however, current repressive actions in tackling violations are not as harsh as before. In addition to the use of softer measures such as advising, educating and surveillance, female officers have been increasingly employed to implement the policing of public space as they appear to be friendlier than their male counterparts.

The 'Interim measure for street vendors' management' policy was then issued in 2009 to include street vendors in urban spaces. The authority promised to establish 120 permitted vending places providing secure and hygienic infrastructure within three years. As Figure 10.2 shows, there are now 65 permitted vending places with 29,515m^2 established areas and 3,596 stalls in use. Street vendors are free to apply for a stall at appointed sites from the government.

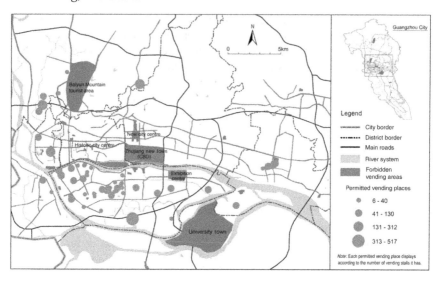

Figure 10.2 The excluded and included geographies of street vendors in Guangzhou.

Source: Drawn by the authors according to the data provided by Guangzhou Urban Management Committee (http://www.gzcgw.gov.cn).

Once permitted, they must sell from the fixed stall, pay management fees and operate regularly as formal merchants. The 'cat and mouse' relation between urban managers and street vendors has thus been replaced with 'state and market' in these spaces. Some of the vendors welcome this approach because they no longer worry about government harassment despite the imposition of some constraints. However, this inclusiveness is conditioned to give priority to a specific city image and urban order.

The authorised sites are restricted to locations where city landscapes, public space order, formal retail businesses and residential daily life are not affected by vending activities. Moreover, these sites may be demolished for urban construction projects in the future. However, the street vendors' needs are not considered in the process of decision-making regarding the organisation of vending spaces. As a result, most authorised sites are located in peripheral or invisible areas such as villages-in-the-city, backstreets in old communities and rural–urban interfaces, which are not as profitable as expected. With doubts about the nature of the inclusionary measure, therefore, many vendors choose to sell in the red areas and continue to challenge the exclusionary practices. Therefore, this inclusiveness does not endow street vendors with the right to sell, but rather the temporary, partial right to sell in given localities. It enables the local state to address the livelihoods of the poor, while securing prime urban spaces. With more green areas established, one might expect more street vendors to be integrated into the formal system. However, it by no means suggests an end to the struggle given the nature of inclusiveness. Rather, new spatial practices by street vendors have emerged due to the ambivalence around the production of exclusionary and inclusionary spaces.

Conclusion

Urban revanchism will not go unchallenged, but invites various counter responses in practice. From the perspective of focusing on the roles of counter forces to revanchism, this chapter presents theoretical and empirical support to assist the understanding of ambivalent policies towards the urban poor. It has explored how some of the counter forces render the exclusionary practices of urban politics in China unstable, difficult to sustain and changeable. It finds that the landscape of exclusionary urban policies was reshaped by the excluded group and a progressive political climate questioning the exclusionary framework. As we have shown, street vendors, as the targeted group the policy seeks to exclude, are able to mobilise various survival strategies that exhaust the authorities and render their exclusionary strategies ineffective. The policy is finally tempered and undermined by the discourse of social harmony at national level. Hence, the existence of the agency of the subordinated in surviving structural constraints, and the benevolent role of the state, results in instability and difficulty for punitive urban policies. This opens up the possibility for the emergence of supportive approaches within spaces of revanchism. Nevertheless, our case does not show an overturn of the punitive framework, but confirms an ambivalent or a post-revanchist approach as recognised in the literature.

The exclusion of street vendors in Guangzhou represents a form of revanchism because it is driven by the NSC campaign that seeks to erase the undesirable from public spaces to create a good city image and enhance urban competitiveness. This exclusion is not driven by the intention to reverse economic recession or dystopian images of urban decline, as suggested by the revanchism thesis based on Western cities. Rather, it is driven by the desires of local politicians to accelerate economic growth and accumulate political achievements. The economic and political desires are, respectively, spurred by the established goal of growth first and the top-down cadre promotion system in China. The *hukou* system, which creates a binary population-division – as well as ethnic and racial factors – supports the revanchist condemnation of street vendors, who are perceived as 'outsiders' in the city. Hence, though outright vengeance is not detected, the exclusion of street vendors represents the essence of revanchism, connoting the intention of the ruling or upper classes to remove the undesirable.

However, revanchist practices such as the NSC campaign, ignore the agency of the social groups they seek to exclude. We have shown the resistance of street vendors is driven by the necessity to survive, or, specifically, their natural consciousness in maintaining the existing livelihood strategy that enables them to adapt to the harsh circumstances resulting from the particular socio-economic restructuring. This necessity underpins their resistance to state repression, creating instability to exclusionary street-vending politics in current Chinese cities. Therefore, even if collective action in China is not as possible as in Latin America (Cross, 1998a), even without associational power vendors develop a flexible, individualised and small-scale activism. This involves strategies of escape, bribery and violently confronting state power to retain the right to use public spaces. Social networks based on *laoxiang* relationships are mobilised to support these actions. Their resistance renders the exclusionary policy difficult to continue as originally intended, and produces a public critique against it. Nevertheless, a limitation of this resistance lies in the fact that it is an autonomous act for surviving exclusionary practice without change to the underlying policy. Thus it is not a deliberate political act, but a moral one. As shown in Creswell's transgression thesis (Creswell, 1996), this form of resistance is limited by its inability to deconstruct spaces of the dominant power. This limit explains why the role of the central state matters in the policy transformation in our study.

China is a centralised country, and the transformation of urban policies cannot be fully understood without taking into account the influence of the national political climate. Despite the devolution of power in the regulation of street vendors, the exclusionary street-vending policy is found to be mediated and reshaped by the discourse of national harmony. In Guangzhou, this process is driven by the responses of local authorities to the issues of the vendors' livelihoods, mitigation of their resistance and harmonious urban management, with the aim of reaching a 'harmonious society'. In addition to capital accumulation, the Chinese state has been increasingly concerned with social needs in order to maintain social and political stability. This concern is reflected in the decision to build a 'harmonious society', and consolidated recently in the proposal of 'inclusive growth'[5] in the 12th Five-Year Plan (2011–15). Nevertheless, social harmony does not mean the goal of growth first is being given up, but connotes the attempt of the central state to maintain a balance between capitalist accumulation and socialist values (Nonini, 2008), and between neoliberal urbanism and social stability (He and Wu, 2008). Therefore, local authorities did not choose to abandon the exclusionary policy, but to adopt an ambivalent one by mixing inclusionary and exclusionary practices in a geographical territory. This ambivalence reflects the intention of the local state to mediate the inherent contradiction between the aim of creating good city images for capital investment, and the need to address the livelihoods and interests of street vendors. This emerging policy is similar to the post-revanchist strategy adopted for San Francisco's homelessness that enables the state to mobilise space while addressing the livelihoods of the poor (Murphy, 2009).

However, because securing prime city spaces rather than truly solving the problems is the goal of a post-revanchist approach, new practices of resistance by street vendors continue in Guangzhou. Retaining social consensus and stability are an important task of the regime, and the Chinese state does not always serve as 'the handmaiden of capital', regardless of the interests and resistances of the subordinated, but promptly tackles the conflict between neoliberal practices and social resistance. It is argued that the landscape of street-vending policy in China is largely shaped and reshaped by the tension between the politics of capital accumulation, vendors' resistance and the state's increasing concerns on social needs. Therefore, this study suggests, a dynamic or historical perspective is essential for understanding the nature of urban political responses to the subordinated. It calls for a research on counter forces at local and supra-local levels in examining the unfolding of punitive/revanchist urban policies, which might be tentative and changeable. In particular, it is worth exploring forms of resistance adopted by local people in surviving, negotiating and challenging political practices in neoliberal cities. Hence, contrasted with the examination of contextualised forms of revanchist politics, this perspective focuses on the driving forces for change, the directions they move towards and the emergence of alternatives that will be more inclusive for the subordinated.

Acknowledgements

The research for this chapter was supported by the National Natural Sciences Foundation of China (Ref: 41130747; 41401169). The usual disclaimers apply.

The chapter is reproduced with permission from John Wiley & Sons. It first appeared as Huang, G., Xue, D. and Li, Z.-G. (2014) From revanchism to ambivalence: the changing politics of street vending in Guangzhou, *Antipode*, 46(1): 170–89. © 2013 The Authors. Antipode © 2013 Antipode Foundation Ltd.

Notes

1 A similar project regarding urban landscapes and environment is the National Civilised City campaign, generally pushed forward after the success in the NSC campaign.
2 For example, 'Guangzhou Building NSC Planning' (Sui Fu [1992] No.24), 'Guangzhou City Reform Proposal for Building NSC' (Sui Wen [2006] No. 37) and 'Guangzhou Work Standards for Building NSC' (Sui Chuang Wei [2007] No. 57).
3 Guangzhou City Landscape and Sanitation Regulation (2007 version) states: 'vendors, who occupy public space and affect city landscapes and sanitation, are subject to be corrected; if refusing the correction, they are subject to a fine of 50–200 yuan'.
4 Cited from the most popular online forum in mainland China – Tianya Community Forum: http://bbs.city.tianya.cn, accessed 5 March 2012.
5 Inclusive growth means that everyone should benefit from economic development, and reflects the attempt of the Chinese state to take social justice seriously in addition to economic growth.

References

Aalbers, M.B. (2010) The revanchist renewal of yesterday's city of tomorrow, *Antipode*, 43(5): 1696–724.
Atkinson, R. (2003) Domestication by cappuccino or a revenge or urban space? Control and empowerment in the management of public spaces, *Urban Studies*, 40(9): 1829–43.
Bannan, R. (1992) Little China: street vending in the free market (1989), *Journal of Developing Societies*, 8(2): 147–59.

Bayat, A. (1996) Un-civil society: the politics of the 'informal people', *Third World Quarterly*, 18(1): 53–72.

Bayat, A. (2000) From 'dangerous classes' to 'quiet rebels': politics of the urban subaltern in the global south, *International Sociology*, 15(3): 533–57.

Brenner, N. and Theodore, N. (2002) Cities and the geographies of 'actually existing neoliberalism', *Antipode*, 34(3): 349–79.

Bromley, R.D.F. and Mackie, P.K. (2009) Displacement and the new spaces for informal trade in the Latin American city centre, *Urban Studies*, 46(7): 1485–506.

Chen, Y.-B. and He, W.-X. (2008) *Why China Is So Successful (1978–2008)?*, Beijing: China Citic Press (in Chinese).

Creswell, T. (1996) *In Place/Out of Place: Geography, Ideology, and Transgression*, Minneapolis: University of Minnesota Press: 163–76.

Cross, J.C. (1998a) Co-optation, competition, and resistance: state and street vendors in Mexico City, *Latin American Perspectives*, 25(2): 41–61.

Cross, J.C. (1998b) *Informal Politics: Street Vendors and the State in Mexico City*, Stanford, CA: Stanford University Press: 40–60.

Crossa, V. (2008) Resisting the entrepreneurial city: street vendors' struggle in Mexico City's historic center, *International Journal of Urban and Regional Research*, 33(1): 43–63.

DeVerteuil, G. (2006) The local state and homeless shelters: beyond revanchism?, *Cities*, 23(2): 109–20.

DeVerteuil, G., May, J. and von Mahs, J. (2009a) Complexity not collapse: recasting the geographies of homelessness in a 'punitive' age, *Progress in Human Geography*, 33(5): 646–66.

DeVerteuil, G., Marr, M. and Snow, D. (2009b) Any space left? Homeless resistance by place-type in Los Angeles County, *Urban Geography*, 30(6): 633–51.

Gardiner, M.E. (2000) *Critiques of Everyday Life*, London: Routledge: 168–74.

GMG (2003) Guangzhou NSC Plan, formulated by GMG (Guangzhou Municipal Government), in *Guangzhou Gazette*, No. 21: 8–41.

Guangzhou UM, Liwan District Branch (2010) Guangzhou urban management law enforcement: problems and solutions, http://www.gzcg.gov.cn/portal/site/site/portal/gzcgw/index.jsp, accessed March 2012 (in Chinese).

Guangzhou Urban Management Committee (2009) Treat stubborn disease by setting permitted space, promote social harmony by mixing inclusion and exclusion, *Guangzhou Urban Management Bulletin*, No. 14 (in Chinese).

Harvey, D. (2005) *A Brief History of Neoliberalism*, Oxford: Oxford University Press: 120–51.

Harvey, D. (2007) Neoliberalism and the city, *Studies in Social Justice*, 1(1): 1–13.

He, S.-J. and Wu, F.-L. (2009) China's emerging neoliberal urbanism: perspectives from urban redevelopment, *Antipode*, 41(2): 282–304.

Huang, G., Xue, D. and Li, Z.-G. (2014) From revanchism to ambivalence: the changing politics of street vending in Guangzhou, *Antipode*, 46(1): 170–89.

Johnsen, S. and Fitzpatrick, S. (2010) Revanchist sanitisation or coercive care? The use of enforcement to combat begging, street drinking and rough sleeping in England, *Urban Studies*, 47(8): 1703–23.

Katz, C. (2004) *Growing Up Global: Economic Restructuring and Children's Everyday Lives*, Minneapolis: University of Minnesota Press: 241–57.

Kerkvliet, B.J.T. (2009) Everyday politics in peasant societies (and ours), *Journal of Peasant Studies*, 36(1): 227–43.

Laurenson, P. and Collins, D. (2007) Beyond punitive regulation? New Zealand local governments' responses to homelessness, *Antipode*, 39(4): 649–67.

Liew, L. (2005) China's engagement with neo-liberalism: path dependency, geography and Party self-reinvention, *Journal of Development Studies*, 41(2): 331–52.

Lin, S.-S., Dai, F. and Shi, H.-P. (2005) *Planning Guangzhou*, 6, Beijing: China Architecture Industry Press: 44–6.

Luo, X.-S. and Huang, M.-C. (2011) Reshape the image of urban management with scientific outlook on development, http://www.gdaiguo.com/sxzg/201102/t20110215_141200.htm, accessed March 2012 (in Chinese).

Mackie, P.K., Bromley, Rosemary D.F. and Brown, A. (2014) Informal traders and the battlegrounds of revanchism in Cusco, Peru, *International Journal of Urban and Regional Research*, 38(5): 1884–903.

Macleod, G. (2002) From urban entrepreneurialism to a 'Revanchist City'? On the spatial injustices of Glasgow's renaissance, *Antipode*, 34(3): 602–24.

McCann, E.J. (1999) Race, protest, and public space: contextualizing Lefebvre in the US city, *Antipode*, 31(2): 163–84.

Meng (2006), The city being focused on equal dialogue between urban managers and street vendors, *New Express*, December, http://www.ycwb.com/xkb/2006-12/20/content_1323078.htm, accessed June 2014 (in Chinese).

Mitchell, D. (1997) The annihilation of space by law: the roots and implications of anti-homeless laws in the United States, *Antipode*, 29(3): 303–35.

Murphy, S. (2009) 'Compassionate' strategies of managing homelessness: post-revanchist geographies in San Francisco, *Antipode*, 41(2): 305–25.

Nonini, D.M. (2008) Is China becoming neoliberal?, *Critique of Anthropology*, 28(2): 145–76.

Papayanis, M.A. (2000) Sex and the revanchist city: zoning out pornography in New York, *Environment and Planning D: Society and Space*, 18(3): 341–53.

Popke, E.J. and Ballard, R. (2004) Dislocating modernity: identity, space and representations of street trade in Durban, South Africa, *Geoforum*, 35(1): 99–110.

Schinkel, W. and van den Berg, M. (2011) City of exception: the Dutch revanchist city and the urban *Homo Sacer*, *Antipode*, 43(5): 1911–38.

Smith, N. (1996) *The New Urban Frontier: Gentrification and the Revanchist City*, London: Routledge.

Smith, N. (1998) Giuliani time: the revanchist 1990s, *Social Text*, 16(4): 1–20

Smith, N. (2001) Global social cleansing: postliberal revanchism and the export of zero tolerance, *Social Justice*, 28(3): 68–77.

Swanson, K. (2007) Revanchist urbanism heads south: the regulation of indigenous beggars and street vendors in Ecuador, *Antipode*, 39(4): 708–28.

Uitermark, J. and Duyvendak, J.W. (2008) Civilising the city: populism and revanchist urbanism in Rotterdam, *Urban Studies*, 45(7): 1485–503.

Van Eijk, G. (2010) Exclusionary policies are not just about the 'Neoliberal City': a critique of theories of urban revanchism and the case of Rotterdam, *International Journal of Urban and Regional Research*, 34(4): 820–34.

Wang, G.-Z., Zhang, B.S. and Zhao R.Z. (1993) *The Study of Economic, Social and Cultural Development in Pearl River Delta*, Shanghai: Shanghai People Press (in Chinese).

Wang, J. (2000) The Mayor's talk on the position of Guangzhou in the future development, http://www.people.com.cn/GB/channel4/985/20000906/219391.html, accessed 8 May 2012.

Wu, F.-L. (2002) China's changing urban governance in the transition towards a more market-oriented economy, *Urban Studies*, 39(7): 1071–93.

Wu, F.-L. (2003) Globalization, place promotion and urban development in Shanghai, *Journal of Urban Affairs*, 25(1): 55–78.

Yangcheng Evening News (2009) Tolerating street vendors is like the situation of 'standing to speak without waist pain', 9 May (in Chinese).

Zhang, Y.Z. (2006) The city being focused on equal dialogue between urban managers and street vendors, *New Express*, December, http://www.ycwb.com/xkb/2006-12/20/content_1323078.htm, accessed June 2014 (in Chinese).

11 Commerce of the street in Sénégal

Between illegality and tolerance

Ibrahima Dankoco and Alison Brown

Summary

In November 2007 three days of demonstrations and riots broke out in response to a government order to evict traders from congested streets of central Dakar. The response was a period of accommodation, leading to the emergence a fascinating process of negotiation between vendor organisations and the city government.

Introduction

In November 2007, the president issued an instruction to the Governor of Dakar region to tidy the streets, interpreted as an instruction to evict. Clearances took place on 21 November; for three days vendors rioted and burnt tyres, closing the city centre. Realising their power to disrupt the city, the president ordered negotiations with the vendors, provided that they were organised. This ushered in a period of negotiation between the state, local government and the vendors that has led to innovations in the accommodation of vendors in the city, but legalisation of the trade remains a distant objective.

Issues of poverty reduction are central to debates on economic growth for both developed and developing countries. In Africa development agendas must clearly address the informal economy, specifically street trade, which is often the main economic activity of the urban poor. The paradox is that, although national governments are willing to address poverty reduction, the informal economy is not addressed and, in practice, street vendors face daily and ongoing harassment by urban authorities.

This chapter analyses the regulatory framework for street trading in Sénégal, examining the land laws and trading regulations which make legal trading almost impossible, and the tension between political tolerance and administrative crackdown that has characterised municipal policy towards street trading over two decades. The chapter first gives a brief historical analysis of the official position on street trading, before analysing the legal framework of street trade, and responses of traders, making recommendations for future reform.

The chapter is based on a literature review and newspaper analysis, a legal review and fieldwork in Dakar in 2010–13 based on extended interviews in six markets with extensive key informant interviews. The findings demonstrate the unpredictability and poverty impacts of the current policy vacuum, the need for an urgent legal review of laws affecting street trade, and innovative and inclusive policy development. In the final stages of the research, the lead author brought together five vendor associations to develop a collective platform to lobby for accommodation and legal reform.

Sénégal: history and context

Street trade has been a traditional activity in Sénégal for many generations, but political and economic policy has set the scene for a rapid increase in recent years. Sénégal won independence in 1960 and is now one of the most stable countries in West Africa. In 1966, President Leopold Senghor's socialist Progressive Union assumed one-party rule, with multi-party politics reintroduced in 1978. Economic growth was affected by drought in 1973–4, and a structural adjustment programme was imposed from 1986–90. In the 2000 election, the second president, Abdou Diouf, was succeeded by opposition leader Abdoulaye Wade of the *Parti Démocratique Sénégalaise* (PDS), who was in turn succeeded by Macky Sall of *L'*Alliance pour la République in 2012 (BBC, 2015). Today, Sénégal's population is around 15 million with a gross national income (GNI) of US$1,050 per capita (2015 estimates); 34% of the population is below the poverty line (at US$1.25 per day), and the level of urbanisation is around 43% (2010–14 estimates) (WB, 2015). A peace accord was finally agreed with separatists in Casamance in 2014.

Emergence of street trading in modern Sénégal

In the years after independence, import controls and trade regulations meant that French capital dominated the formal sector in trade and industry, with some businesses still run by companies of Lebanese origin (Thioub *et al.*, 1998). Dakar Chamber of Commerce represented these formal businesses. African businesses linked to the ruling party gradually won control over the imports, and the Islamic brotherhoods, which by the end of the nineteenth century controlled the lucrative peanut trade, took advantage of contraband routes to the Gambia (Brown *et al.,* 2015; Brown *et al.*, 2010). This paved the way for the emergence of a flourishing of informal economy, largely tolerated by the government as a means of providing jobs and cheap consumer goods during the periods of high inflation and economic recession in the 1970s and 1980s (Thioub *et al.*, 1998).

Immediately after independence, the economy was largely agricultural and people managed to find work without too many difficulties. However, the first global oil shock of 1974 coincided with a long period of drought, resulting in a rapid drop in farm incomes and employment, which created pressure for informal sector work. Thus, from the 1970s onwards several factors contributed to a rapid increase in street trade:

- the drought and subsequent rural exodus to cities: droughts from 1968 to the mid-1970s, and again in the mid-1980s, led to famine and rural dislocation on a massive scale, particularly from the dry Sahelian regions of north-east Sénégal;
- the Structural Adjustment Programme adopted in 1985 at the instigation of the World Bank and IMF led to massive unemployment and a shift from formal to informal work;
- associated changes to agricultural policy and the Structural Adjustment Programme failed to produce expected results, which exacerbated rural poverty, increasing the rural exodus to the cities;
- the opening of the country to people from the sub-region, which led to young people from Mali, Sierra Leone and Guinea arriving *en masse* – many resorting to street trade;
- the arrival of Chinese traders in large numbers, facilitating access to low-value goods for sale, allowing young people to bypass the traditional trading sectors dominated by the *mouride* and other brotherhoods.

Structural adjustment was particularly destructive. Between 1986 and 1990, neoliberal economic reforms led to widespread privatisation of state industries and bitter fights to control the lucrative import trades (ibid.).

In Sénégal, distinction is usually made between three types of vendors: *tabliers*, who trade from a kiosk, table or fixed site; *ambulants* or hawkers; and *semi-ambulants*, who use a vehicle – cart, trolley or bicycle – and settle in one place for short periods.[1] A wide range of consumer goods is sold, the main categories being: fruit and vegetables; fish; clothing and textiles; sports clothing; second-hand goods; traditional medicines; bags, shoes and leatherwear; households goods; mobile phone accessories; pre-paid mobile credit; sale of Touba spiced coffee, toothpicks, hair plaiting supplies; DVDs and cassettes; plastic bags; sunglasses; jewellery; watches; canned food, Chinese imports; toiletries; perfumes; street tailors service; hairdressing, etc.[2]

Economic logic suggests that street vendors need to be accessible to an influx of potential customers based on the 'proximity principle', so the busiest locations

Figure 11.1 Sénégal has a long history of street vending, but infrastructure is poor.

for street trading are in the central business district (CBD), and in and around various large markets and bus stations on the peninsular. Attempts to relocate vendors to remote trading sites are thus unlikely to be successful, which suggests some scepticism about the resettlement solutions proposed in Dakar by political and municipal authorities. Legality/illegality does not appear to affect choice of location.

Formerly a stepping stone to commerce, where a merchant's children received schooling in trade,[3] street trade is now a permanent activity for many vendors, and there are limited opportunities for young street vendors to graduate into more secure occupations (Lyons *et al.*, 2008). The social and ethnic background of street vendors has now diversified to include all social classes, including the indigenous Lebou population of the Cap-Vert peninsular, whose children are going into street trade (key informant interview). The growth of street trading and congestion of the sector poses significant urban management problems.

Politicisation of street trade

For over two decades, street trade in Dakar has been highly political. In 1989 an attempt by the government to increase customs duties and value-added sales tax (TVA) was foiled when businesses protested, closing shops and offices and paralysing the CBD. In 1990 the *Union national des commerçants et industriels du Sénégal* (UNACOIS) was set up, representing large and small-scale vendors, including the informal sector. Operating through strikes and direct action, particularly in 1993 and 1996, UNACOIS successfully fought off increases in import taxes and attempts to apply TVA to street vendors. By 1998 UNACOIS claimed a membership of 70,000, including many small-scale vendors and hawkers (Thioub *et al.*, 1998; Brown *et al.*, 2015).

During the 2000 election, President Wade courted vendors and, after this, merchants gained confidence, stopping the president in a cavalcade on the way from the airport; he agreed that vendors should not be evicted. As the Secretary General of the *Groupement national des jeunes marchands de Dakar* said in an interview,

> *A partir de 2000 on a noté un véritable changement d'attitude. Le président de la république a été interpellé par les ambulants et le cortège présidentiel s'est arrêté pour permettre au Président de parler aux ambulants porteurs de brassards rouges. Ces derniers se sont plaints des tracasseries, le Président les a rassurés et depuis ils ont commencé à devenir des tabliers dans presque toutes les rues de Dakar.*
>
> (Since 2000 there has been a real change in attitude. The president was questioned by *ambulants* and the presidential motorcade stopped to allow the president to speak to those holding red armbands. They complained of harassment, the president reassured them and since then they started setting up kiosks in almost all the streets of Dakar.)

This was a significant event, as many hawkers then built kiosks along the roads and an influx of new vendors occurred.

L'image du vendeur ambulant s'est améliorée: tout le monde veut devenir ambulant. La stabilité est plus grande dans le secteur et les ambulants sont plus sécurisés. Pour peur ils seront formalisés dans leur lieu d'exercice. On trouve même de nos jours des tables sur l'avenue Ponty avec comme charge seulement une taxe de FCFA 150 par jour et tout cela en toute sécurité. C'est extraordinaire!

Vendor association representative

The image of vendors has improved and everyone wants to become a vendor. Stability in the sector has increased and vendors are more secure. Out of fear they have become formalised at their trading site; we even find tables along the Avenue Ponty which are only charged a daily tax of FCFA 150, which grants them security. It's amazing!

In 2007, the *Ministère du commerce* took a census of street vendors in Dakar, recording around 8,000 vendors, thought to have increased to 15,000 by 2010. Some vendors' groups disputed these figures, which seemed an under-count if the membership figures for *Synergie des marchands dits ambulants pour le développement* (SYMAD) and the *Fédération des associations des marchands tabliers du Sénégal* (FAMATS) were correct. This highlights a perennial problem, the difficulty of estimating the scale and economic contribution of street trading. The census reported that 91% of vendors were men, 82% were under the age of 30 and 80% were funded by credit from suppliers, confirming earlier findings (Lyons and Snoxell, 2005; Brown *et al.*, 2010). An international symposium on formalising street trade was arranged with the International Labour Organization (*Bureau International du Travail*, ILO) and Senegalese government in 2008.

During the clear-up in November 2007, many vendors had no advance warning of the evictions, although UNACOIS was informed. After the riots, and with perhaps a *volte face*, the president dismissed the governor and announced that negotiations would take place with the vendors, provided that they were organised. A consultative commission was set up with the *Ministère du commerce*, and the *Mairie de Dakar* set up a weekly consultation group with vendors' representatives (key informant interviews). After '*les émeutes*' (evictions) of 2007, the minister of small and medium-sized enterprises was also dismissed (key informant interviews).

Several new vendors' organisations were then formed in 2009, including SYMAD, comprising 12 associations with a combined membership around 7,000, and FAMATS, with 15 associations and over 5,400 members (key informant interviews). SYMAD proposed that street trading be legalised and managed at street level by vendor committees (key informant interviews).

Meanwhile, in a shift from its previous militancy, UNACOIS has shown distinct reluctance to support street vendors, a sector which is essentially illegal, in 2010 reportedly making a statement of support for the evictions to the *mairie* (key informant interviews). UNACOIS sought to restrict the influx of vendors during public holidays such as the Muslim festival, Tabaski, but also supported a 2011 strike against an increase in the daily toll, so perhaps the attitude is pragmatic.

In the decade following the 2000 presidential election, there was unprecedented political acceptance of street trade, and the political support of young street vendors was courted.

Street trade and the law in Sénégal

The legal and regulatory situation in Sénégal situates the relationship between street vendors and the state between the illegality and tolerance in defiance of the law. There is a repressive legal arsenal which is sometimes used instead of tackling the economic poverty that underlies street trade. Thus, street trading is criminalised and the security forces track and penalise vendors on a daily basis. Official reaction is to confiscate the goods of vendors, which are left to spoil, or to take action through the courts.

Sénégalese law on street trading was strongly influenced, first, by the socialist agenda of President Senghor (1960–80) and, second, by economic liberalism (from 1990s). Externally, trade regulations attempted to address the requirements of globalisation and World Trade Organization (WTO) rules. These interventions tried to strengthen urban development strategies as a response to development needs. Three distinct phases in the evolution of street vending legislation can be identified:

- *Centralised state socialism, 1960–80s*: During the socialist period the country was under control of the powerful UPS-PS political party (*l'Union progressiste sénégalaise – Parti Socialiste*). Freedom of expression was non-existent and all popular media were under state control.
- *Democratic deepening and economic liberalisation of the 1990s*: From the 1980s, since multi-party politics was introduced, the country has experienced political turmoil and economic crises that led to massive unemployment and unprecedented urban growth. Legislation of this period was intended to safeguard urban areas and streets against the rural–urban influx.
- *Post-2000 liberal policies*: These are epitomised in the 2001 Constitution (amended in 2009), which suggests that, at national level, there appears to be a general willingness to resolve the problems of street vendors.

Sections 8 and 25 of the national Constitution establish the right to work:

- s.8: The Republic of Sénégal guarantees all citizens fundamental freedoms, economic, social and collective rights. These rights and freedoms include. . . the right to work
- s.25: Everyone has the right to work and the right to claim employment. No one shall be discriminated against in their work due to their ethnicity, sex,

> opinions, choices or political beliefs. Workers can join a union and defend
> their right through trade union action

but these provisions are far from being implemented, and vendors remain vulnerable to shifts in policy.

Trading regulations

Both trading law and land law, and their political context, have implications for street vendors in Sénégal, discussed in the next two sections. Legislation governing street trading is primarily regulated by *Loi 67-50* of 29 November 1967 that covers trading in the public highway and in public space, as amended by *Loi 75-105* of 2 December 1975. This was further amended by *Loi 76-018* of 6 January 1976, relating to the sale of goods on the highway and in public space, and a decree, *Décret 87-817* of 25 June 1987. This legislation controls trading from a fixed location.

Article 4 of the decree states that: 'No one may exercise the profession of *tablier* (selling from a table) without authorisation from the Minister of Commerce. Permitted zones for setting up stalls are fixed by order of the regional governor' and that: 'authorisation results in the issuance of a business card of the form annexed to this order and must be presented on demand'. Article 6 states that: 'it is also prohibited to solicit clients on the road and in public places not designated by order of the Regional Governor'. This clause appears to target street vendors without explicitly naming them.

Loi 76-018 required *tabliers* to have a business licence, but excludes mention of hawking: the practice of hawking is thus, at best, tolerated. *Décret 87-818* modifies *Loi 76-018*, which under Article 2 forbids, '*toute vente ambulante sur la voie et dans les lieux publics, même de façon occasionnelle des produits et denrées visés à l'article premier*' (all hawking in public places, including the products and food noted in Article 1), strengthening restrictions on sale of goods on the highway.

During the 1990s, the strict legal regime was somewhat relaxed, as the aim was to adopt an urban management approach that was closer to citizens' needs. In parallel, a programme of local government decentralisation was adopted, setting up a system of *regions, communes* (districts/municipalities) and *communautés rurales* (CR) (rural communities) – also known as *collectivités locales* (CT) (rural collectives) – with competences defined in legislation and budgetary powers to raise funds through taxes, levies, property income, fees and user charges (Gilbert and Taurgourdeau, 2012).

Thus, in practice, most street trade is now regulated by local government. *Loi 96-06* of 22 March 1996 establishes the local authority code and powers, and *Loi 96-07* of 22 March 1996 establishes the decentralisation agenda, defining the competencies of local authority in urban management and urban planning, zoning for commercial and other activities and the management of space.

As the neoliberal agenda took hold, street trading also became subject to trade legislation – for example:

- *Loi 94-63* of 22 August 1994 laid the foundation for liberalising the economy, regulating prices, competition and trade disputes; the law liberalises trade and gives the opportunity for all citizens to be involved in trade.
- *Loi 94-68* of 22 August 1994 set up measures to safeguard national production from illicit commercial practices.
- Loi 94-69 of 22 August 1994 regulates economic activity and permits buying and selling property, except where public interest is infringed. Article 3 requires that: '*Les professionnels non soumis à la procédure de l'autorisation doivent faire une déclaration de leurs activités conformément aux dispositions organisant le registre du commerce ou le registre des métiers*' (Professionals without authorisation must make a declaration that their activities are consistent with the requirements of the Register of Commerce or the Register of Occupations).
- *Décret 95-817* of 20 September 1995 gives responsibilities to the Ministry of Commerce to implement national trade policy, undertake market surveillance and promote consumer protection.
- *Décret 2008-514* on the quality of life and consumer standards.

In the late 1990s Sénégal adopted two laws (*Loi 96-06* and *Loi 96-07*) on the transfer of state powers to the CRs. An interview with a senior member of staff at the *mairie de Dakar* confirmed that:

> *C'est le Code des collectivités locales qui traite de ces choses, on y parle de l'OVP (occupation de la voie publique). C'est le maire qui autorise l'occupation de cette voie publique et le préfet est informé. Les collecteurs empochent l'argent et ne versent rien dans les caisses. Et de toutes les façons, ce qu'ils collectent est minime par rapport au budget de la mairie.*
>
> Key informant interview
>
> It is the local government code that deals with these matters, including occupation of the highway. The mayor authorises occupations, and the prefect is informed. However, fee-collectors are pocketing the money and pay nothing into the coffers, and what they collect is minimal compared to the cost to the *mairie.*

Thus controls over the management of space and cost controls are often at odds with the objectives of freedom of trade, expressed in *Loi 94-63.*

From 2000 onwards the political regime was relatively supportive of street trading. In November 2007, following evictions from central Sandaga and rioting in the city centre, there was a national focus on the problems of the young and unemployed. This led to promulgation of *Décret 2009-1410* of

23 December 2009, setting up *La Commission nationale d'assistance aux jeunes marchands* (National Commission for the Support of Young Vendors, CONAJEM).

However, all the legal texts that regulate street trade exclude *ambulants* (mobile vendors or hawkers), and *tabliers* (fixed vendors) are subject to a condition that they must have authorisation and a business registration card. Thus the practice of street trade is, at best, tolerated and, at worst, castigated.

Land law

Land law, similarly, went through phases of regulation closely related to the political and economic philosophy of the period. For many years Sénégal has had a dual system of land rights combining private and customary title. Legislation derived from the pre-independence French Civil Code still defines the legal system of land registration. The land registration was first introduced under colonial rule in 1906, and modified in 1932. A decree in 1925 sought to facilitate the registration of customary title, but failed because it granted individual rights, not collective rights more suited to African culture at the time.

Shortly before independence, a lands study recognised the problem of parallel property regimes and proposed a constitutional guarantee of property rights. In the first post-independence socialist period, land titles were constituted under *Loi no. 64-46 relative au Domaine National* of 17 June 1964, amended by *Loi 76-66* of 2 July 1976, still the main law governing land rights, which reflected the ideology of Sénégalese socialism and sought to harmonise formal and customary rights. This considered national domain as '*un droit de synthèse original poursuivant deux objectifs essentiels: la socialisation de la propriété foncière plus conforme à la tradition négro-africaine et le développement économique du pays*' (an original synthesis of law pursuing two main objectives: social ownership of land in keeping with African tradition and the country's need for economic development) (Sow, 1997 : 1).

To justify the 1964 Act, President Senghor argued that '*de revenir du droit romain au droit négro-africain, de la conception bourgeoise de la propriété foncière à la conception socialiste qui est celle de l'Afrique Noire traditionnelle*' (in examining the link between Roman law and African land rights, the socialist conception of land rights is that of traditional Black Africa) (ibid.).

The Act decreed that all land belongs to the state, with the exception of land with formal registered title. The law recognises three broad categories of land rights: *state domain*; *private domain* regulated by Decree in 1932 and registered in the *Livre Foncier* (Land Record); and *national domain*, which covers all non-registered land – divided into urban zones, classified zones (forestry or protected areas), *zones de terroirs* (village land) and pioneer zones (undefined areas) (Sow, 1997; Hesseling, 2009).

Three categories of land rights were prescribed. The *titre foncier* is a registered title which enables land to be bought and sold; the *permis d'occuper* is an occupancy right which can be revoked at any time by the state and the *lease* retains land ownership with the state but allows use of land on a rental basis (Dankoco, 2011). Village land is allocated to those who demonstrate productive use (*mise en valeur*).

Land legislation was further reformed in 1976 to cover land expropriation and compensation. Article 1 of *Loi 76-67* states that: '*L'expropriation pour cause d'utilité publique est la procédure par laquelle l'Etat peut, dans un but d'utilité publique et sous réserve d'une juste et préalable indemnité, contraindre toute personne à lui céder la propriété d'un immeuble ou d'un droit réel immobilier*' (Expropriation for public good is a procedure through which the state can, subject to a public purpose and prior reservation of just and prior compensation, compel any person to cede land and property, or any effective property right). National domain lands cannot be registered except by the state, a strategic move that allows the state gradually to access land rights. According to Sow (1997), this creeping privatisation of national domain land is contrary to the spirit of the *Loi 64-46*.

Loi 87-11 of 24 February 1987 and its implementing *Décret 87-271* of 3 March 1987 authorises the sale of public land for housing in urban areas, thus reducing the amount of inalienable rural land, free from appropriation. Once in urban land use, the land rights pertain to those who have been allotted land.

The last land reform, in 2011, radically changed Senegalese land tenure, but does little to help street vendors. Recommendations from the 2006 national land commission resulted in *Loi 12-2010* on property registration, but the law excludes property rights for urban land used for small-scale business and trade.

Décret 2009-1302 of 20 November 2009 establishes rules for the operation of *L'agence nationale de l'aménagement du territoire* (National Spatial Planning Agency), whose mission statement set out in Article 3 includes '*lutter contre les encombrements divers de la voire publique (marchands ambulants non autorisés, véhicules abandonnés ...)*' (fighting against congestion on the public highway (from unauthorised street vendors, abandoned vehicles. ..)). However, vendors and artisans operating from the highway and sidewalks are legitimate economic actors (e.g. mechanics, carpenters, shoemakers, etc.). Nowhere in the new laws is there any express recognition of the need to trade.

The last reform, adopted 22 February 2011, radically changed the Senegalese tenure. These two laws, *Loi 12-2010* on the system of land ownership and *Loi 11-2010* on permits for residential use and security for land titles, guarantee the rights of property holders, who can then use the land as collateral for a mortgage or loan.

Nevertheless, after more than 50 years of implementation, the enactment of the national domain Act is still partial (Dankoco, 2011; Benkhala and Seck, 2011; Boye, 1978). There are, thus, no formal property rights for street vendors, but a thriving informal land market exists in urban and peri-urban areas for housing land and agreements over use of national domain land are sometimes ratified by rural councils. It is likely that socially accepted and informal rights also extend to trading sites.

Urban management paradigms post-2007

After the riots of 2007, street vendors again claimed the headlines, but tension ensued. While the *communues* and CTs were seeking to find a negotiated settlement with street vendors, national spatial planning policy remained restrictive.

Nevertheless, the riots created political space for dialogue, and the authorities adopted practical steps to deal with street vendors. Faced with the prospect of presidential elections in 2012, the state adopted the policy of 'divide-and-rule' to take advantage of the power relations between the state, *mairie* and vendor organisations, but – under pressure from multilaterals promoting the poverty-reduction agenda – some compromise was inevitable. A coordinated approach was adopted that included:

- clearance of street vendors from the downtown area;
- creation, in 2010, of CONAJEM (*Décret 2009-2014*);
- creation, in 2011, of a national agency for street vendors, *l'Agence nationale d'appui aux marchands ambulants* (ANAMA); this was cancelled in 2014 and amalgamated into *l'Agence nationale pour la promotion de l'emploi des jeunes* (ANPEJ) (Apanews, 2014); and
- the organisation of a presidential council with street vendors.

Progress was also made on the ground. The *mairie de Dakar* held weekly meetings with vendor organisations and eventually purchased a site in rue Félix Eboué costing FCFA 2 billion, designed as a design-build-and-operate contract, with vendors' associations involved in the selection of contractors. New sites are also being sought for displaced vendors in Parcelles Assainies, Thiaroye and Mbeubeuss. In autumn 2010 the President declared construction of a new market at Pètersen, and at the *Conseil des Ministres* in February 2011 he announced the creation of a new agency charged with addressing the issue of the *marchands ambulants*, to include relocation and kiosk canteen projects (key informant interviews).

The urban development strategy for Dakar, 2025, aims to make the city a modern metropolis. This was adopted under *Décret 2009-622* of 30 June 2009, which approved implementation of the *plan directeur d'urbanisme de Dakar*. The plan is reviewed regularly to take account of demographic, social, economic and environmental change. However, within this context street vending is seen as contrary to the image of this modernity, and the new sites are not steeped in trading customs as are the older markets of Dakar (Marfang, 2015). The elections of February 2012 divided street vendors, some of whom supported Abdoulaye Wade in his bid for a third presidential term and others who showed allegiance to his ultimate successor, Macky Sall. Meanwhile, street clearances continued in preparation for the summit of Francophone heads of state in October 2014.

At Céntenaire, an interview site located north of the Grande Mosquée de Dakar, just north of the CBD, conflict was evident between four key groups of protagonists: the *mairie*, residents, young job-seekers from the interior and

Chinese traders, now a substantial trading bloc in the city (ibid.). This long dual-carriageway road is now thronged with young Senegalese traders selling imported goods from the adjacent Chinese-run shops that line the road. Evictions were again threatened in 2013 and vendors, for a long time accustomed to the protection of Wade, thought they would not take place, but they were finally evicted as part of preparations for Independence Day parades on 4 April 2014.

At Pètersen, in June 2015, the market by the bus termimus was burnt down, allegedly by the *mairie*. Riots ensued and many vendors were reportedly in despair. '*Tout est parti en fumée. C'est la mairie qui a ruiné nos cantines. Dans quel pays sommes-nous? Comment peut-il nous priver de notre gagne-pain à quelques jours du Ramadan,*' one vendor said (Sakho, 2015). (Everything is burnt. The council has ruined our kiosks. What kind of country is this? How can they deprive us of our livelihood during Ramadan?)

Finally, in July 2015, the new Félix Eboué market was complete, offering sites for 3,000 vendors. The council held an open day and invited vendors to apply for places by the deadline of July 2015, with priority for vendors listed in 2010. Vendors, however, complained that the rental was too high and have refused to move in (Traoré, 2015). However, the numbers accommodated will not satisfy demand in the city centre; meanwhile, young traders are still being cleared from the streets, and a solution is far from being found. Trading remains illegal across a whole raft of legislation. Vendors claim rights through direct action, but official policy seems to vacillate between negotiation and repressive control.

Experience from the streets

Vendor interviews were carried out in six markets in Dakar with different characteristics:

- *Sanadaga* – formerly the main municipal market and its surrounds in the heart of the CBD (built in 1933; in August 2013, after a flood, the market temporarily closed for safety reasons; it was later gutted by fire);
- *Pètersen* – an informal market near the main bus terminal, selling mainly Chinese imports;
- *HLM* – a market in a planned suburb, selling fabric, fashion and electrical goods;
- *Céntenaire* – on a dual carriageway north of the CBD, selling mainly Chinese imported goods;
- *Tilène* – a local market west of the CBD; and
- *Grand Yof* – a low-income suburban market.

This section reports on two aspects of the survey that indicate rights claim – the payment of daily fees, and the experience of evictions and harassment as an indication of contested rights. Of the 143 vendors interviewed, 83% were men and 17% were women, with a higher proportion of women vendors in HLM (12/30). Almost all vendors were working six to seven days a week and 12 hours a day, indicating that this was their main employment.

Payment of fees suggests a degree of acceptance of vendors' legitimacy, and this was evident in all six markets. Of 142 valid responses, 90% (128 vendors) paid daily fees, all except two to the municipality, mostly at a rate of FCFA 100–300 a day – one said that he ran away to avoid paying and another that he had the protection of a Marabout (holy man), so did not pay. Rates of payment were similar across the six markets.

Despite the political gains of the last decade, insecurity was common. Among vendors almost all (90% of 138 valid responses) had at some time experienced one or more of the four main dimensions of insecurity: harassment, confiscations, fines and evictions. Harassment (59% of 137 responses) and confiscations (59% of 138 responses) were commonly reported, with about a third of vendors having been fined (32% of 137 responses) and a third evicted (34% of 137 responses). Some had been sent to jail for trading in the road reserve. There were, however, significant differences between markets, with Sanadaga and Cénentaire reporting much higher rates of harassment (75% of 24 responses and 68% of 22 responses respectively) than the other markets. Although the disaggregated sample size is small, the experience appears to be widespread, as the following quotes suggest. While '*les émeutes de 2007*' still loomed large in vendors' experience, a crucial problem is unpredictability.

> *Plusieurs fois on m'attaqué – une fois même on m'a pris mes bagages et j'ai laissé là bas pour aller cultiver, jusqu'à la fin de l'hivernage. C'est après que je suis venue récupérer mes biens en payant une amende de FCFA 3.000. Si on me traque ce jour là, je ne vends rien et c'est difficile de subvenir à mes besoins.*
>
> Male vendor, HLM, selling children's shoes
>
> Several times I was attacked, and one time someone even took my bags, so I left there to go back to farming until the end of the winter. Afterwards, I came to recover my property by paying a fine of FCFA 3,000. If you follow me since that day, it's been difficult to support myself.

> *C'était à la veille d'une fête je me suis installée devant l'emplacement de ma mère quand une personne en civil est venu me demander de ranger et m'a insulté. Je lui a insulté et il m'a amené à la police après j'ai payé FCFA 7.000 pour récupérer mes biens.*
>
> Female vendor, HLM, selling scarves
>
> On the eve of a festival I installed myself in front of my mother's space, when a person came and asked me to tidy up and insulted me. I insulted him back, and he took me to the police after which I paid CFCA 7,000 to recover my property.

La veille de la Tabaski 2009 à Sicap Mbao, j'étais vendeur de couteau. Je vendais dans le foirail quand un policier en civil est venu m'interpelé pour me dire qu'il est interdit de vendre des armes blanches car c'est dangereux. Lorsqu'il m'a amené à la police ou je suis resté une heure avant de reprendre mes biens. J'avais payé CFCA 3.000 d'amende.

Male vendor, HLM, selling calculators
and nail clippers

On the eve of the Tabaski in 2009, at Sicap Mbao, I was a knife salesman. I sold at festival sites when a plainclothes policeman hassled me and said that it is illegal to sell knives because it is dangerous. Then he took me to the police station and I stayed an hour before reclaiming my goods. I paid a fine of CFCA 3,000.

Il y' a un jour la police est venue dans la matinée nous traquer, et je n'avais pas de quoi payer l'amende – alors ils m'ont référé (au Tribunal). Une fois au Tribunal on nous demande de ne plus aller vendre sur la voie publique, et puis ils nous relâchent.

Male vendor, Sandaga, selling
second-hand clothes for men

One day the police came in the morning hassle us, and I had nothing to pay the fine – then they referred me to (the Tribunal). Once at the Tribunal we were asked not to sell on the street, and then they released us.

Vendors were starting to recognise that collective action to achieve recognition as legitimate urban actors was key and, as a result of the research, the lead author set up a working group of the main street vendor organisations. Previously, the groups had very different objectives, some working through social networks and others more familiar with official processes, so the meetings were important in building bridges. Four groups came together in a series of meetings: SYMAD, FAMAT, *Groupement National des Jeunes Marchands de Dakar* (GNJMD) and the *Syndicat des vendeurs du marché du poisson*. Two were members of the global network StreetNet International.

Vendors' primary aim was to become accepted by the state and *mairie* as legitimate operators, leading to a participatory process for reviewing and improving the regulations. Their core demands were as follows:

- acceptance by the authorities that public space may be allocated for trade under certain conditions;
- designation of zones of economic activity in urban plans and inclusion of regulations to address the problems of street vendors;
- a change to the legal and regulatory framework governing the public realm (including land and business regulation) to recognise the legitimacy of street trade;
- establishment of a consultative process for addressing the problems of street vendors in Sénégal.

Conclusion: the need for legal and conceptual reform

Despite the gains, street vendors in Sénégal suffer continued harassment and threat of eviction because of their lack of recognition in the law, particularly resulting from the bans on trading in public space and along the public highway, and the lack of social protection. There are also effects on the country's economic growth as those vulnerable to eviction have little chance for economic accumulation. There are also recurring problems in residential neighbourhoods because of tensions between street vendors and residents.

Sénégalese law remains anachronistic and does not recognise the economic realities of the day. This research demonstrated that vendors claim rights, but are subject to a conflicting legal and institutional domains. Urban management problems arise because of conflicting jurisdictions over urban space, with many institutions involved, including: the *Ministère du commerce, les mairies de ville* on trade management, the *Ministère des Finances* on taxation and land rates, the *Ministère de l'Environnement* in environmental protection, the *Ministère de la Petites et Moyennes Entreprises* (small and medium-sized enterprises) responsible for the informal economy, the *Ministère de l'intérieur* on security and the public highway, the *Ministère de la décentralisation et des collectivités locales* on local government and the *Ministère de la Justice* on access to the law.

Repressive laws still exist and can be invoked at any time to harass vendors. The associations of formal businesses do not maintain good relations with vendor organisations, and local governments do not understand appropriate mechanisms to resolve problems of congestion and trade management. A fundamental legal review is essential to accommodate this urban reality that cannot now be denied. From the research carried out on this project, five key areas for reform emerge:

1) *business law*: the blanket ban on hawking as a micro-enterprise should be revoked, and hawking by individuals must be legalised, in line with other legal businesses;
2) *planning and zoning laws*: the zoning and design of public space in cities must be reviewed to develop models that facilitate hawking and vending alongside other urban functions;
3) *land law*: legal instruments must be reviewed to create modern yet viable tenure for vending in public space;
4) *obligations of hawkers*: as legal actors, vendors must fulfil their obligations to the state, including management of public space and paying taxes;
5) *institutional overlaps*: regulation of street vending must be simplified, and participatory planning processes introduced.

With the foundation of democracy in Sénégal and expansion of a free press, street vendors are increasingly able to stake their claims and articulate a vision. The process of collective organisation, originally encouraged under the Wade government, has enabled vendor associations to become more sophisticated in their approach. As Moustapha Mbaye, leader of the *Groupement national des jeunes marchands de Dakar*, said,

> *Aujourd'hui, on ne se plaint plus de reconnaissance car le Président parle de nous dans son discours de fin d'année (31 décembre 2011) et nous participons à des conseils présidentiels qui nous sont dédiés. Notre projet aujourd'hui c'est de faire une pétition pour faire abroger la loi sur la vente de rue.*
>
> Today we are no longer asking for recognition, as the president mentioned us in his end-of-year speech, and has invited us to participate in a presidential council dedicated to vendors. Our plan today is to make a petition to repeal the law on street vending.

However, while high-level political support secured a degree of stability for vendors, ongoing harassment of vendors from low-level bureaucrats is rife. Over a period of 25 years, rights have been established through a process of direct action which creates space for political negotiation. However, dialogue is not enough and, without legal status, vendors remain vulnerable to regime change and political whim.

Meanwhile, the illegal character of street trade means that any gains will be tentative. What is essential now is formal recognised provision that guarantees street vendors their rights as conferred by the Constitution (specifically the right to work). Beyond mere recognition, there is a need for a national policy on street vending in Sénégal, building on the experience of India and elsewhere (see Chapter 8). The space for dialogue can be increased by strengthening the capacity of street vendor organisations to defend their interests, and the development of street vending legislation that sets out processes to achieve compromise between the rights of all actors with legitimate interests in the public domain of the street – including government, vendors, pedestrians and others – to re-envisage the street as *un lieu d'exercice du commerce* (a commercial domain).

Notes

1 There is currently a debate about the meaning of different terms used for street trading. Representatives of SYMAD, a well-known street vendors' organisation, argue that *ambulants* should be called 'itinerant merchants' rather than hawkers, to avoid problems of legal recognition.
2 This typology of activities was established at a workshop at the Faculty of Economics and Management, Université Cheikh Anta Diop, with four vendors' associations.
3 President of the management committee, Sandaga market, formerly the largest permanent market Dakar.

References

Apanews (2014) Sénégal: l'agence nationale pour la promotion de l'emploi des jeunes officiellement lancée, Seneweb.com, http://www.seneweb.com/news/Societe/senegal-l-agence-nationale-pour-la-promotion-de-l-emploi-des-jeunes-officiellement-lancee_n_133779.html, accessed May 2016.

BBC (2015) Senegal profile: a chronology of events, http://www.bbc.co.uk/news/world-africa-14093813, accessed September 2012.

Benkahla, A. and Seck, M. (2011) Pour une véritable concertation sur les enjeux et objectifs d'une réforme foncière au Sénégal. IPAR (Initiative Prosective Agricole et Rurale), http://www.inter-reseaux.org/IMG/pdf/Policy_brief_-_reforme_fonciere-4.pdf, accessed May 2016.

Boye, A.-K. (1978) Le regime foncier sénégalais, *Ethiopiques* (14), http://ethiopiques.refer.sn/spip.php?article645, accessed May 2016.

Brown, A. (2006) Social, economic and political influences on the informal sector in Ghana, Lesotho, Nepal and Tanzania, in Brown, A. (ed.), *Contested Space: Street Trading, Public Space and Livelihoods in Developing Cities*, Rugby: ITDG.

Brown, A., Lyons, M. and Dankoco, I. (2010) Street-traders and the emerging spaces for urban citizenship and voice in African cities, *Urban Studies*, 47(3): 666–87.

Brown, A., Msoka, C. and Dankoco, I. (2015) A refugee in my own country: evictions or property rights in the urban informal economy?, *Urban Studies*, 52(12): 2234–49.

Dankoco, I.S. (2011) Le commerce de rue Au Sénégal: entre illégalité et tolérance, unpublished research report.

Gilbert, G. and Taurgourdeau, E. (2012) The local government financing system in Senegal, in *The Political Economy of Decentralization in Sub-Saharan Africa*: 207–64.

Hesseling, G. (2009) Land reform in Senegal: l'Histoire se répète?, in Ubink, J., Hoekema, A. and Assies, W., *Legalising Land Rights: Local Practices, State Responses and Tenure Security in Africa, Asia and Latin America*, Leiden, Netherlands: Leiden University Press: 243–71.

Lyons, M. and Snoxell, S. (2005) Sustainable urban livelihoods and market-place social capital: a comparative study of West African traders, *Urban Studies*, 42(8): 1301–20.

Lyons, M., Dankoco, I.S. and Snoxell, S. (2008) Capital social et moyens d'existence durable: quelle stratégie de 'survie' chez les commerçants urbains du Ghana et du Sénégal?, *Revue Ouest Africaine de sciences économiques et de gestion*, 1(1): 12–37.

Marfang, L. (2015) Dakar 2025: L'avenir du commerce ambulant face aux stratégies d'aménagement de la municipalité, *Metropolitiques*, http://www.metropolitiques.eu/Dakar-2025-L-avenir-du-commerce.html, accessed August 2015.

Sakho, B. (2015) Face-à-face ambulants-mairie et policiers: Samedi de feu à Petersen, http://www.lequotidien.sn/index.php/component/k2/face-a-face-ambulants-mairie-et-policiers-samedi-de-feu-a-petersen, accessed August 2015.

Sow A.S. (1997) Domaine National, la Loi et le Projet de Réforme, in : La Revue du Conseil Economique et Social, 2, February–April, http://unpan1.un.org/intradoc/groups/public/documents/idep/unpan004225.pdf, accessed August 2015.

Thioub, I., Diop, M.-C. and Boone, C. (1998) Economic liberalisation in Senegal: shifting politics of indigenous business interests, *African Studies Review*, 41(2): 63–89.

Traoré, H. (2015) Recasement des marchands ambulants, Le mairie de Dakar livre un centre commercial de 3,000 places, *Enquête*, http://www.enqueteplus.com/content/recasement-des-marchands-ambulants-la-mairie-de-dakar-livre-un-centre-commercial-de-3-000, accessed August 2015.

WB (World Bank) (2015) Sénégal data, http://data.worldbank.org/country/senegal#cp_wdi, accessed August 2015.

Main legislation and regulations affecting street traders

Loi no. 64-46 du 17 juin 1964 relative au Domaine National.

Loi 67-50 du 29 novembre 1967 relative à la réglementation des activités qui s'exercent sur la voie et dans les lieux publics.

Loi 75-105 du 20 décembre 1975 modifiant et complétant la loi n° 67-50 du 29 novembre 1967 relative à la réglementation des activités qui s'exercent sur la voie et dans les lieux publics.

Décret 76-018 du 6 janvier 1976 réglementant la vente sur la voie et dans les lieux publics.

Loi 76-66 du 2 juillet 1976 portant Code du domaine de l'Etat.

Loi 87-11 du 24 février 1987 autorisant la vente des terrains domaniaux à usage d'habitation situés en zone urbaine.

Décret 87-271 du 3 mars 1987 sur l'application des lois n° 87-11 du 24 février 1987.

Décret 87-817 du 25 juin 1987 modifiant certaines dispositions du Décret n° 76-018 du 6 janvier 1976.

Loi 94-63 du 22 août 1994 sur les prix, la concurrence et le contentieux économique.

Loi 94-68 du 22 août 1994 relative aux mesures de sauvegarde de la production nationale contre les pratiques commerciales illicites.

Loi 94-69 du 22 août 1994 fixant le régime d'exercice des activités économiques.

Décret 95-817 du 20 septembre 1995.

Loi 96-06 du 22 mars 1996 portant code des collectivités locales.

Loi 96-07 du 22 mars 1996 portant transfert de compétences aux régions, aux communes et aux communautés rurales, organisant la décentralisation.

Décret 2008-514 du 20 mai 2008 portant création et fixant les règles d'organisation et de fonctionnement de l'Agence nationale du cadre de vie et de la qualité de la consommation.

Décret 2009-622 du 30 juin 2009 approuvant et rendant exécutoire le plan directeur d'urbanisme de Dakar.

Décret 2009-1410 du 23 décembre 2009, portant création de la commission nationale d'assistance aux jeunes marchands (CONAJEM).

Loi 12-2010 portant régime de la propriété foncière.

12 The politics and regulation of street trading in Dar es Salaam

Colman Msoka and Tulia Ackson

Summary

Protests and clashes between street traders and police in Tanzanian cities are not new. Since the *nguvukazi* hawker's licence was cancelled in 2003, making hawking effectively illegal, official practice has vacillated between tolerance in periods close elections and repression after elections, and street vendors have defended their claim through direct action. This chapter explores the fluctuating political rhetoric around street trading in Tanzania, and demonstrates how the legislation and bylaws impose requirements that make compliance for street traders unattainable.

Introduction

In March 2006, a letter from the Prime Minister's office issued a government order requiring local authorities to end their tolerance of 'informal petty trade'. Evictions took place in all large urban centres, including Dar es Salaam, Arusha, Mbeya, Morogoro and Mwanza. The order resulted in widespread unrest in which 'many people were … injured before anti-riot police quelled the violence with tear gas' and two died; the government then imposed a six-month delay to allow traders to prepare, but the evictions eventually proceeded (BBC, 2006; Ntetema, 2006). Up to 1 million street vendors were thought to have lost their place of work through eviction or intimidation (Lyons and Msoka, 2010), and in Temeke District alone, one of Dar es Salaam's three municipalities, prosecutions were issued against 40,000 street vendors, from an estimated population of 250,000 traders (Lyons and Msoka, 2008).

Since then periodic 'clean-ups' have resulted in further unrest. In Iringa in May 2013, street vendors flouted eviction orders and rioted, leading to the arrest of an opposition MP and 60 street vendors (Sabahi, 2013), and in June 2014 local newspapers reported that the Prime Minister was 'striving to intervene and control the *machinga* (street trader) crisis, that was reportedly threatening people's lives and their property in the lakeside city of Mwanza' (*TDN*, 2014). Many other such incidences between vendors and the police occurred.

This chapter draws on field research in Dar es Salaam and a legal review undertaken under the three-year Economic and Social Research Council (ESRC)/ Department for International Development (DFID) study on law, rights and regulation for the informal econmy, based on an historical review, and detailed legal analysis. The chapter is divided into three main sections. After the introduction, the second section examines the political economy of street trading and policy evolution in Dar es Salaam and the third section sets out the legal framework for street vending, proposing that the 'right to vend' – within reasonable limits or constraints – should be considered a basic economic right. The conclusion in the final section draws together the debate.

Defining street vending

Street vendors embrace a distinct group of people working in the informal sector which accommodates all activities outside the regulatory framework, including hawking, undeclared domestic work, bartering and petty trade. Although the name connotes the use of 'street' as the main place of vending and therefore earning income, street vendors, as the ILO argues, include the informal workforce operating on the 'streets, sidewalks, public parks, outside any enclosed premise or covered workspace' (ILO, 2002: 49; Bernabe, 2002). Thus street vending includes:

> not only those street vendors who sell goods but also a broader range of street workers who sell services and produce or repair goods, such as: hairdressers or barbers; shoe shiners and shoe repairers; car window cleaners; tailors specializing in mending; bicycle, motorcycle, van, and truck mechanics; furniture makers; metal workers; garbage pickers and waste recyclers; headloaders and cart pullers; wandering minstrels, magicians, acrobats, and jugglers; beggars and mendicants ... street vending is a large and diverse activity: from high-income vendors who sell luxury goods at flea markets to low-income vendors who sell fruits and vegetables alongside city streets.
>
> ILO, 2002: 49

In Tanzania, street vendors include those selling goods or offering services in public spaces, hawkers, peddlers and petty traders, including all the categories of informal sector working by the road side or near designated formal markets, irrespective of their 'occupations' such as fruit or vegetable traders, furniture makers, hairdressers and barbers. Drawing comparisons with work elsewhere, street vendors in Tanzania are likely to account for about 12–24% of the informal sector, embracing all types of people – youths, men and women – the latter predominating in the sector (WIEGO, 2016).

Although all Tanzania cities have street vendors, this study focuses Dar es Salaam as Tanzania's commercial capital, with the largest number of street vendors, and where the impacts of different policy agendas can be seen. After relocation of street vendors to the famous Machinga Complex, a purpose-built but poorly located market, many returned to the streets to find customers; likewise many who moved to organised markets also returned to their old spots in the streets (e.g. moving from Urafiki to Buguruni, then from Buguruni to Mabibo and Kimara Mwisho to Mbezi Mwisho). Another reason for selecting Dar es Salaam is that street vendors are frequently at loggerheads with the City Police (militia). Often, the approach to dealing with vendors in the city is later applied to other major cities in the country. This study examines the socio-economic and political challenges of street vending in Dar es Salaam and their rights under the existing laws of Tanzania.

Political economy of street vending in Tanzania

Although street trading in Dar es Salaam existed in colonial times it was heavily restricted, and confined to the African settlements of Kariakoo, Ilala and Temeke (Lugalla, 1997). Informal sector traders mainly sold food, fruits and vegetables; clothing was mainly sold by Arabs – *guo guo*; later some Africans began to join the sector, mainly after independence (Jecha and Mdundo, 1999; Lugalla, 1997). Rural–urban migration was very limited and those who contravened the law were charged, so the sector remained small. Rural–urban migration was restricted as the colonial authorities wanted to maintain agricultural production, and cyclical migration was common, where people would work on the land during the farming period during and after harvest, and work in towns at other times (Owens, 2010). After Tanzanian independence in 1961 four main phases of evolution can be identified.

Early post-independence and ujamaa

The *ujamaa* policy, African socialism and self-reliance, introduced at the Arusha party conference in 1967 (Nyerere, 1977), led to widespread economic transformation and nationalisation of large agriculture and industrial projects. By the late 1970s the policy had largely failed, due to external factors such as the 1970s oil crisis, global collapse of coffee prices, Tanzania's 1978–9 war with Uganda and internal factors such as mismanagement of the newly nationalised industries by inexperienced professionals (Brown, 2006; Owens, 2010). The challenges came at time when the country had started an ambitious industrialisation strategy where new industries employing hundreds of people were also failing. These failures had several other implications, including unemployment from closing factories and lack of markets for agricultural crops, which led to wider economic crisis in the country (Kulaba, 1982).

To cope with the economic crisis, many rural residents flocked to the cities, especially during non-farming seasons, often to engage in petty trading. By the second half of the 1970s, street trading in Dar es Salaam had become a political issue for both local and central government authorities. In the mid-1970s, the government rounded up street vendors in Dar es Salaam and took them to *ujamaa* villages for communal production of agricultural goods (Osafo-Kwaako, 2011).

The government was spending significant resources developing the rural sector, by constructing roads, schools and health centres (Nyerere, 1982). Furthermore, the policy priority of the time had a rural bias and thus urban areas were not getting the required attention. With increasing population it was thus becoming more of a problem to the government (Lugalla, 1997).

Both government and urban residents saw rural migrants in Dar es Salaam as pursuing activities that were either economically unnecessary or illegal. In 1976 and 1983 the government organised repatriation programmes, resettling migrants to their villages of origin. Another strategy used in the early 1980s in Dar es Salaam required families to register for food rations in a government-sanctioned store, *duka la kaya* (Tegambwage, 1985), where the number of people who lived in the household was registered as well. This was a move to make sure that only those who had a reason to lived in the city, and the rest returned to the villages for farm work. Neither policy significantly halted the migration.

Rural–urban migration was also stimulated by the inadequate level of formal sector wages, so employees set up vending businesses to cushion the cost of urban living and brought in rural family members to help run businesses (Bagachwa, 1994; Msoka, 2005). Dar es Salaam is colloquially known as *bongoland*. *Bongo* is a Kiswahili word for brain, referring to the period when people had to be creative to generate extra income in the city.

International migrants too played a role, including Asians brought in as cheap labour by European settlers, who pioneered trades such as tailoring, tin-smithing, grocery stores and food processing. Indian traders were very successful, establishing schools, hospitals and places of worship (Mbilinyi, 1974). Among the poorest migrants were the Yemenis, who were often hawkers (Jecha and Mdundo, 1999).

1980s: economic hardships and the rise of street vending

The development of mass informal markets in urban Tanzania has its beginnings in the country's economic hardships of the 1970s and 1980s (Kahama *et al.*, 1986; Campbell and Stein, 1992; Young, 1983). Following oil price shocks of the mid-1970s, and a fall in agricultural product prices in global markets, Tanzania's (centrally planned) economy was badly shaken (Horne, 1987; Miti, 1987). Imports at the time were by state-owned firms as the country's private sector was limited. Foreign currency was rationed and priority was given to sectors considered important. This had an impact on new investments and job creation and the economy was not growing fast enough to absorb the growing labour force. This provided an opportunity for the informal sector to fill the gap (Burton, 2007).

According to Gibbon (1993) and Kahama *et al.* (1986), Tanzania found itself in crisis following the collapse of the then East African Community (EAC) in 1977. The EAC brought together the five countries of Burundi, Kenya, Rwanda, Tanzania and Uganda, and had a number of jointly owned and operated corporations for post, air transport, railways and harbours, customs, research and metrological services. New institutions were needed to fill the gap caused by the collapse, and resources were taken from other development priorities. The collapse of the EAC thus crippled economic growth, leading to yet more unemployment.

Relations between Tanzania and Uganda had been strained for several years and broke out into war in 1978 which lasted for 18 months (ibid.). The direct cost of the war, which was unbudgeted, had devastating effects on the economy of Tanzania. The indirect effects, such as need for resources to rebuild bombed infrastructure and destroyed farms, added substantially to the war costs.

After the war Tanzanians went through a difficult time, popularly known as *halingumu*. Imports were controlled and the little foreign currency available was used to import oil, medicine and some intermediate goods (Wagao, 1990). The control and rationing of foreign exchange left firms and manufacturers without crucial inputs or parts for their machines. The situation created scope for the emergence of a *parallel market* in the country, conducted mostly in the form of street-level vending and backyard sales (Maliyamkono and Bagachwa, 1990; Ndanshau and Mvungi, 2002; Gibbon, 1993).

Meanwhile, in 1983 the Penal Code, Cap 16, was enacted, branding all self-employed people and street vendors as 'unproductive, idle and disorderly', justi-fied on the basis that street trading was a subversive activity that challenged socialist principles (Burton, 2007). The government also cracked down on traders who withheld stock to vend informally, such as employees in state-owned firms and parastatals who dominated import sectors – for example, regional trading companies, agricultural supplies companies and cooperatives (Tegambwage, 1985). The shortage of goods in the economy was felt in all aspects, and the government lifted a ban on importing second-hand items. A large second-hand business sector emerged selling *mitumba* (any used item from abroad, e.g. cars, clothes, shoes, sandals, linens, curtains, handbags, or furniture), allowing second-hand goods to replace the shortages. Sweatshops and home-based enterprises also thrived, such as oil-seed processing, bakeries and artisanal activities of carving, pottery, batik making and basket weaving (Tripp, 1996, 1997; Maliyamkono and Bagachwa, 1990). Vendors were no longer vendors of fruit and vegetables, but of imported manufactured goods. The liberalisation of the economy gradually led to both old and new companies facing competition and wishing to maximise their returns – some private companies also employed young men to vend old stock outside stores to attract customers.

Thus, after the economic policy reforms of the mid-1980s, Tanzania was inte-grated into the global capitalist system, imports and exports were deregulated, the economy was liberalised and the private sector began to feature strongly. The informal sector was permitted because the government was not able to create enough jobs for the unemployed and could not pay adequate wages to its employ-ees. The sector was thus promoted as a way of cushioning inflation and supple-menting what was not available. During this period the streets of Kariakoo, the main trading and commuter interchange area in the city, saw lines of vendors trading openly. These changes were initiated by President Mwinyi (1985–95), whose *laissez-faire* leadership and popular slogan, '*ruksa*' (permitted), allowed people to fill in what the state could not do (Sanders, 2001).

Government response: the *nguvukazi* licence

The government response to the rise of street vending in the mid-1970s to early 1980s was to round up unemployed youths and send them to the countryside (Tripp, 1996). In areas surrounding Dar es Salaam – for example, Makongo, Mbezi, Chanika, Rumo, Mpiji, Gezaulole, Kibugumo, Mwanadilatu and Kigamboni – people were allocated agricultural activities. However, many returned to the city within a short time (Burton, 2007), which led to the promulgation of the Human Resources Deployment Act, Act 3, 1983 (HRDA), popularly known as the *nguvukazi act*. Under the HRDA, every able-bodied person was required to have a legal job and had to be known by authorities in their respective areas. In Dar es Salaam, individuals with jobs that did not qualify were questioned and repatriated to rural areas for agricultural work. The Act also made special provision the disabled, youths and women.

The 1983 the Regional Commissioner of Dar es Salaam, Mr John Mhavile, listed the following activities as illegal: shoe shining, food vending, selling local brew, knife sharpening, car washing in non-sanctioned areas and all kinds of street hawking activities (Msoka, 2005). Legal petty businesses for operators with a licence included: newspaper vending, shoe repair, hair dressing, tailoring, woodwork, masonry, electrical and electronic technicians and music recording. In broad terms, hawking was banned and vendors were encouraged by the act, if not pushed, to settle and do business in known locations. This had unintended consequences in that it encouraged rather than discouraged petty trading.

In essence, the HRDA contributed to the growth of street vending and hawking, as, by paying fees to the council, traders were, in a way, legalised. Over the years, local authorities added different activities, with the strategy of increasing revenue through HRDA licences, and the Act proved to be a major source of government revenue.[1] HDRA (*nguvukazi*) licences were meant to cater for small-scale traders who were unable to rent formal business premises. However, the HRDA was passed when the country was in a severe economic crisis, following a failed *coup d'état*, and when the country was reintroducing local authorities ten years after they were abolished. During this period, most of the public corporations were in crisis and some were on government subsidy (Msoka, 2005).

From 1996–2000, to fend off bankruptcy, Dar es Salaam City Council was taken over by the Dar es Salaam City Commission. Recognising the *nguvukazi* programme's potential as a source of revenue, the City Commission expanded licensing requirements beyond kiosks to include street tables and open-air vendors (Mhamba and Titus, 2001). Thus the City Commission's view of street vendors changed from a problem to a resource, which opened space for dialogue with vendors, developed a formal relationship with them by fixing fees (at TSh100) and required them to form associations. However, street vendors were still clearly defined as informal and never given a permanent lease or contract (Tume ya Jiji, 1997).

Figure 12.1 Dar es Salaam: in the late 1990s, traders used portable structures (left); the system is no longer used (right).

Street vendors in the areas of Kariakoo Market, Kisutu Market, Ocean Road Hospital, Zanaki Street and Kitumbini area were required to use special structures or portable shelves for vending (Figure 12.1), and fees were paid on the type of structure and the space, not the volume or value of goods, thus trading again expanded. This strategy was good for both parties – it boosted revenue for the Commission, and for traders it created, for the first time, a space for dialogue with local government, the Office of District Commissioner, the Regional Commissioner and the Prime Minister's Office.

The measures encouraged further growth in the sector and created space for vendors to oppose the Commission. The new associations helped vendors gain power and represented them in meetings with local government. Vendors sometimes pressured the Commission, or brought them grievances with other institutions such as the Ministry of Labour and the Department of Youth. Vendors' associations also used media to resist orders for relocation, demolition or eviction, call for better infrastructure, question the validity of paying daily fees in the light of frequent harassment and demand legal protection.

1990s onwards

Since the 1990s, there have been swings of government policy, moving from guarded recognition to political control (ESRF, 2006; Sayaka, 2006). The Tanzania of today is more capitalist-oriented and the rural bias has waned. This is partly due to the policies of multilateral donors such as the World Bank (WB) and the International Monetary Fund (IMF), which favour urban rather than rural sectors. The government has introduced a number of urban (re)development programmes and projects including:

- 1992–2002: the Sustainable Cities Programme (SCP), a participatory strategic urban development process adopted in nine cities, supported by UN-Habitat (Nnkya, 2006);

- early 1990s: the Urban Sector Engineering Project (USEP), which developed a policy framework for urban management, service delivery and infrastructure investment (WB, 2002);
- 1996–2004: the Urban Sector Rehabilitation Project (USRP), designed to promote sustainable economic development and urban poverty alleviation through basic infrastructure rehabilitation and local government capacity building (WB, 2004);
- late 1990s: the Urban Health Project (UHP) a short-term programme in Dar es Salaam.

Dill, 2007

The government of President Mkapa, 1995–2005, attempted to search for ways of solving the problem of informality by formalising the sector. To do this, the President invited Peruvian economist Hernando de Soto to advise on formalisation of the informal sector (Lyons and Msoka, 2008). The plan was to help informal sector micro-traders make the transition to formalisation to advance their business. A large-scale business formalisation programme was set up under its acronym MKURABITA (Kiswahkili, 2008). Although the programme still exists, it has not been able to address the needs of the poor and micro-traders as anticipated, partly due to the lack of policy focus on informal sector micro-traders.

Under the government of President Kikwete, elected in 2005, a letter from the Prime Minister to local urban authorities called for them to recover order in city and demanded that all activities in unauthorised areas be removed. The order led to widespread evictions in which perhaps 1 million vendors lost their livelihoods (Lyons and Msoka, 2010). Yet again, a policy of clearances dominated with brutal effect. In 2011 Dar es Salaam City Council developed a new six-storey Machinga Complex, but it remained empty several years after completion, reflecting a mismatch between its design and the dependence of vendors on passing trade.

In a quiet rebellion, street vendors have again colonised many of the spaces from which they were evicted. Attempts to relocate them have been met with riots and vandalism as a way to stop the authority from pushing them out. At the time of writing, as the 2015 election neared, a coordinated opposition had emerged; the ruling party again sought to avoid controversial action that might alienate the large voter community of vendors. Nevertheless, without recognition as legitimate economic actors, and legal reform to protect their constitutional rights, street vendors remain vulnerable to the eviction and harassment that has characterised relations with the authorities for many years.

Tanzania's legal framework governing street vending

The legal regime for street vendors in Tanzania is largely negative; harassment, evictions, confiscation of goods and arrests of street vendors are sanctioned by law. This section argues that the extent of legislation affecting street vending is contrary to the Constitution and infringes vendors' human rights. The main legislation and its implications for vendors is summarised below.

The United Republic of Tanzania Constitution of 1977

Everyone, including street vendors, is required by the Constitution (URT, 1977) to respect both principal legislation and subsidiary legislation in the form of rules, regulations and bylaws.

Several clauses in the Constitution are relevant. As Tanzanian citizens, street vendors are entitled to protection of their person, dignity and property and guaranteed the right to life, right to work and related rights such as those relating to livelihood and general wellbeing.

Property rights

Art. 24 establishes property rights and provides that:

1) Every person is entitled to own property, and has a right to the protection of his property held in accordance with the law.
2) Subject to the provisions of sub-article 1), it shall be unlawful for any person to be deprived of his property for the purposes of nationalisation or any other purposes without the authority of law which makes provision for fair and adequate compensation.

Property rights refer to both real (landed) and personal (movable) properties. Both customary ownership and intermediate rights of occupancy are protected as property. Street vending mostly deals with personal properties and, therefore, movable articles such as second-hand clothing (*mitumba*) etc. The Constitution also includes a right to fair compensation. The issue is, do street vendors enjoy this right?

Right to a livelihood

Livelihood, which denotes 'a means of securing the necessities of life', is provided for in the Constitution in the form of the right to work and the state's responsibility to protect the people's welfare.

The state's duty towards people's welfare is covered under the 'Fundamental Objectives and Directive Principles of State Policy' in the Constitution, and since wellbeing is measured through people's livelihoods, then the state is obliged to protect street vendors so that the necessities of life are secured. However, the Fundamental Objectives establish the principles on which legislation should be based but are not directly enforceable in the courts (URT, 1977 Art. 7(2)).[2] Thus, although the right to a livelihood is covered under the Constitution, that right is unenforceable. Under the Fundamental Objectives, Art. 9(e) requires the government to ensure that 'every person who is able to work does work'.

However, the Bill of Rights, which is enforceable in the courts, also contains a right to work (Art. 22(1)). Art. 25(1) places a duty on people to work as 'work alone creates the material wealth in society, and is the source of the well-being of the people and the measure of human dignity'. Is the street vendors' right to work protected in actual practice? This is another issue this chapter examines.

Work is also linked to the right to life, and Art. 14 states that 'every person has the right to live and to the protection of his life by the society in accordance with the law'. Thus, street vendors' right to work and duty to engage in lawful and productive activities to earn a living are directly linked to their survival.

Other rights

Apart from rights relating to property and livelihoods, other rights within the Constitution are essential to realising the rights of street vendors. First is the protection of one's person, residences and privacy (Art. 16(1)). Second is the protection against unlawful arrest, as arrest must be under the circumstances and procedures prescribed law (Art. 15(2)). Third is the principle non-discrimination, and a right to equality before the law (Art. 13). However, these provisions are yet to be invoked in courts of law; only then can the Constitutional rights be confirmed (Art. 64(5)).

Land Laws 1999

In 1999, Tanzania enacted two land laws: the *Land Act, No. 4 of 1999* and the *Village Land Act, No. 5 of 1999*. The Land Act governs all the lands except the village land, while the Village Land Act governs village land. In respect of street vendors, the Land Act is particularly relevant, as most street vending is based in urban areas.

The Land Act (URT, 1999) categorises land into three main groups: general land, village land and reserved land (Section 4(4) Cap 113). Reserved land includes land designated under the Urban Planning Act, the Highways Act, the Public Recreation Grounds Act and land reserved for public utilities (Section 6(1) (a) and (c)). Street vending often takes place in areas that are designated as reserved land.

Under s.4(1) of the Land Act, all land in Tanzania is public land entrusted to the president for benefit of all the citizens. The Public Trust Doctrine envisages that all the citizens should have guaranteed access to land, but, in practice, land is not accessible to many groups including street vendors, women, youth and marginalised communities.

The Land Act also requires land to be utilised sustainably, and all persons exercising rights under the Act 'to ensure that land is used productively and that any such use complies with the principles of sustainable development' (Section 3(e)). Thus, sustainable development objectives require street vendors to comply with issues of licences and permits and to conduct their business in permitted areas (formal markets) for orderly urban development.

Planning laws

The planning laws in Tanzania include the *Land Use Planning Act, 2007*, and the *Urban Planning Act, 2007*. These are put in place to deal with sustainable land use in the planning areas.

The *Land Use Planning Act, No. 6 of 2007* (URT, 2007a) was enacted to provide for the preparation, administration and enforcement of land use plans. Among the many objectives this Act, those that are relevant to street vendors include the provisions to 'facilitate the creation of employment opportunities and eradication of poverty'.[3] Thus planning authorities are required to ensure planning activities facilitate rather than hamper the creation of employment opportunities, and eradicate rather than perpetuate poverty. This is not what the street vendors get from the implementation of the Act.

Land use plans do not consider vendors a distinct category of the community which needs to be accommodated in their plans. As such, the actualisation and enforcement of the Land Use Planning Act perpetuates unemployment by denying street vendors space from which to work, and the constant demands of bribes and confiscation of goods increases poverty as street vendors, instead of using their income for their own development, always have to please the law enforcers. This would not be the case if city and municipal plans were in line with Section 3(d), which requires 'needs and aspirations of the various sections of the population' to be taken into account in all land use plans.

Additionally, s.28(1) lists matters to be included in all land use plans, including the preservation of open space and defined paths on the land. As there is no express mention of street vendors' space in the plans, street vendors end up occupying open space, roadsides and paths. Arrests, evictions and confiscation by law enforcers would be considered as implementing the Act, which requires preservation of the space occupied 'illegally' by street vendors. Nevertheless, two paragraphs of the Second Schedule to the Act provide for employment land and strategies for livelihoods to be accommodated within the plan, which could take into account the needs of the informal sector and thus street vendors, but this is not implemented.

The *Urban Planning Act, Act No. 8 of 2007* (URT, 2007b) essentially provides for a comprehensive planning system in urban areas. It restricts doing business without a licence from relevant authorities.

Section 2 defines the term 'street' to includes 'any road, square, footway or passage, whether a thoroughfare or not, over which the public has right of way, and also the way over any public bridge', while a 'public street' means any street over which the public has a right of way and which is or has been usually repaired maintained by the government or local government authority (LGA). This means that streets are usually under control and management of the public authority, either local or central government.

The Urban Planning Act provides for a number of fundamental principles of urban planning such as to facilitate the creation of employment opportunities and eradication of poverty (s.3(c)). Likewise, Section 4(1) provides that the Act is set to facilitate efficient and orderly management of land use and promote sustainable land use and practices. However, the enabling aspect of the legislation is rarely followed. In the words of the ILO:

> Street vendors are often viewed as a nuisance or obstruction to other commerce and the free flow of traffic. Since they typically lack legal status and

recognition, they often experience frequent harassment and evictions from their selling place by local authorities.

ILO, 2002: 49

Under s.7(1) of the Act, planning authorities are tasked with conserving buildings, premises or land and open spaces, to reserve and maintain all land planned for industrial and commercial purposes in accordance with the approved planning schemes (s.28) and to regulate development to conform to such schemes (s.42), but the provisions do not provide for the use of land by the informal sector.

It is submitted that street vending should be an aspect specified within the planning scheme. The role of street vending in job creation and reduction of poverty cannot be overemphasised. Thus, the focus by LGAs on formal markets limits the ambit of the objectives of the Land Use Planning Act and the Urban Planning Act and their ability to contribute to wider planning objectives such as the National Strategy for Growth and Reduction of Poverty (MKUKUTA) and Tanzania's Vision 2025.

Legal grounds for evictions

Evictions of street vendors from their vending stations have legal backing in Tanzania in that land law, planning law, highway law and local government law restrict the use of streets, open space and public areas for business. These laws prohibit conducting business in areas not designated for street vending by planning authorities; vending in areas planned for other things such as public utilities; vending without licences or permits in planned areas; obstruction of free passage of persons and vehicles along the highway, and causing nuisance to others. Street vendors, therefore, are prohibited from all these acts which form the basis for frequent evictions, as outlined below (see also Box 12.1).

The *Land Use Planning Act, 2007*, grants planning authorities powers to control and restrict particular uses in order to ensure orderly development, make bylaws in order to regulate uses and density, and reserve land for open spaces (s.46). Anyone who violates these provisions commits an offence under the Act. The defaulter may be liable to a fine not exceeding TSh 2 million or three years in jail, or both. Street vendors, as their activities take place in public places such as open spaces and parks, thus face risks of arrest, confiscation of goods and payment of fines if they are found guilty of violating the law. The fine is too high for a street vendor to afford.

The *Urban Planning Act, No. 8 of 2007* also grants powers that can be used to evict street vendors. S.29 prohibits any development on land without the planning consent granted by the planning authority and makes planning consent a condition for the issue of licences. It is not common for street vendors to seek 'consent', let alone planning consent, before they start their business. Box 12.2 gives three examples of evictions under planning law.

The *Highways Act, Cap 167*, as amended in 1969, restricts obstructions on the highways. Under s.39, acts that may amount to an offence include anything that:

1) encroaches on any public highway …
2) in any manner, wilfully prevents any person or any vehicle from passing along any public highway;
3) obstructs the free passage on a public highway by exposing goods or merchandise of any description.

Thus street vending, including display of goods or merchandise along the road and putting up a structure for display, is an offence punishable by fine of TSh 500 or a jail term up to three months (s.51). Thus the law is restrictive, but does not show how road reserves may be used for different activities, including small businesses. Under the Highways Act, street vendors may be evicted if they have obstructed the free passage of people or vehicles along the highway. Although the law seems to allow activities on the highway where one has permission, street vendors often trade without permission, and authorities are usually reluctant to issue such permissions.

Box 12.1 2011–12 evictions of street vendors in Dar es Salaam

In January 2011, hundreds of vendors were evicted and their structures destroyed along the Morogoro Road in Dar es Salaam, from Kimara to Magomeni, as they were vending in the road reserve and road widening was proposed to make way for the Dar es Salaam Rapid Transit (DART) project. The evictions took place without warning, and with force (*Mwananchi*, 2011).

In April 2012, hundreds of vendors at the Ubungo Junction, along the Mandela Express Highway and Morogoro Road, were evicted by the Tanzania Electricity Supply Company (TANESCO) on grounds of safety, as they were trading under high-voltage cables, and it was argued that they could contract cancer or fire could erupt at any time. Their part was taken by their elected parliamentarian, who argued that there were around 150,000 petty traders in Dar es Salaam and 3,000 at Ubungo alone. Police and City Militia stormed the area using tear gas and destroying vendors' property and wares, but the vendors gave TANESCO three days to find them another plot (Felister, 2012). In practice, little progress has been made and vendors have returned to the area.

In November 2012, street vendors in Mwanza were evicted from the junction of Pamba Road, MitiMirefu Road and Tanganyika bus stand in Mwanza Central Business District (CBD). Properties were destroyed by the City Militia, one person was shot dead and many others injured in the incident (*Tanzania Daima*, 2012).

The *Local Government (Urban Authorities) Act, 1982, Cap.288*, empowers the Minister to establish an urban authority in a township, municipality or city (s.5). The established authorities are corporate bodies, having perpetual succession and common seal; capable of suing and being sued; and capable of holding, purchasing and disposing of properties (s.14).

Urban authorities' statutory duties include cleansing and markets and the inspection of food and drink. Urban authorities have a duty to seize and destroy all foodstuffs unfit for human consumption; prevent public nuisance, which may harm public health or order; regulate noxious or dangerous trade or businesses; set fees and issue licences or permits to facilitate the regulation of trade or business; and prohibit unauthorised markets other than public markets established by the authority (s.62(1e, i, m)). Street vending is often considered a nuisance and therefore categorised as illegal by the relevant authorities.

S.62(2) and the Schedule (items 34, 35, 36, 85) to the Act provide a list of functions which may be performed by urban authorities, which include: regulating markets, constructing market buildings and regulating trade outside established markets; regulating and controlling the fixing of and collection of storage, rents and tolls in markets; and prohibiting or regulating the use of streets in the area. These provisions are used to promulgate bylaws, further restricting street vending. Urban authorities are also charged with maintaining peace, order and good government, and promoting the welfare and economic wellbeing of everyone within their areas of jurisdiction (s.60(1a,b,c)). S.98 of the Act gives the general penalty for violation of the Act as a fine of up to TSh 50,000 (s.103).

Box 12.2 Recourse to the courts

The use of urban authorities' powers is exemplified by the case of *Republic v. IddiMtegule*, where the Area Commissioner of Dodoma demanded explanation from the Primary Court Magistrate as to why an accused vendor, charged with selling *maandazi* (buns), was released. An order by an Area Commissioner had banned the sale of food in public places to prevent the spread of cholera. Buns were not among the prohibited items. The Area Commissioner was furious and wrote accusing the Magistrate of bias and thwarting attempts to stamp out cholera. The Magistrate informed the Area Commissioner that he was interfering with independence of the judiciary. This case is important in that it shows how urban authorities, without regard to the letter of law, may harass, arrest and 'convict' street vendors who may not have actually breached any law in place.

High Court of Tanzania at Dodoma (PC),
Criminal Session No. 1 of 1979 (unreported) Peter, 1997: 488

Legal processes for eviction

Under the Constitution (Art. 107A), every person who defaults the law must be brought before a court of competent jurisdiction, and basic human rights contained under the Bill of Rights in the Constitution must be observed in the whole process of arresting, prosecuting and sentencing. Art. 13 (6) (b) of the Constitution of URT is very clear that 'no person charged with criminal offence shall be treated as guilty of an offence until proved guilty of that offence'. However, urban authorities may impose fines without charging an offender in a court of law, provided that the offender admits in writing to the offence and an official receipt is issued for the fine paid.

Some cases filed before the Resident Magistrate's Court of Dar es Salaam at Sokoine Drive for violation of the city regulations or municipal bylaws were settled by the payment of fines. Other cases were settled through withdrawal of the local government prosecution or forfeiture of the bond by the accused person for failing to appear in court (see *R v. Charles Phidelis & Others*, Resident Magistrate's Court of Dar es Salaam at Sokoine Drive, Criminal Case No. 295 of 2010, unreported). In other cases street vendors' goods are confiscated or destroyed by order of the court (see *R v. SelemaniYasin & Others*, Resident Magistrate's Court of Dar es Salaam at Sokoine Drive, Criminal Case No. 122 of 2010, unreported). Sentences may include fines, jail or community service.

Human rights implications of evictions

It is clear that the evictions of street vendors abrogate a number of basic rights of those involved, particularly the right to life, right to work, right to own property, right to protection of the acquired property and the presumption of innocence.

The right to life is protected by the Constitution; street vendors are effectively 'condemned to death' when they are evicted from their vending stations – cutting their only means of survival affects their right to life. In some cases, law enforcers have actually caused deaths of street vendors who were objecting to evictions. Evictions also affect the right to work, which is protected by the Constitution. It is submitted that evictions affect the right to work in that affected street vendors will not have a place to vend and therefore will not be able to work.

Evictions also affect street vendors' constitutionally protected property rights, particularly where goods and merchandise are confiscated. A 2015 court case in Durban has recently shown such confiscations to be unconstitutional in South Africa (see Chapter 9). Moreover, evictions of street vendors abrogate their right to be presumed innocent until proven guilty. The actual practice is that the street vendors are condemned as guilty until proven innocent, and during evictions street vendors' merchandise is confiscated and structures are destroyed even before cases are brought to court. At times, law enforcers abuse and beat street vendors, thereby violating their protected human rights and sometimes deaths have been reported such as that in Mwanza (Box 12.1).

Therefore, the manner in which evictions are carried out is highly questionable in terms of whether the human rights enshrined in the Constitution are observed. Beatings of the street vendors, confiscation of their properties without proper procedures or destruction of their property are the order of the day in urban areas in Tanzania.

Municipal-level regulation

Under ss.88 and 89 of the Local Government (Urban Authorities) Act, municipal councils have legislative power to make bylaws to carry out their functions. This section highlights the main bylaws affecting street vending in the three munici-palities in Dar es Salaam, Kinondoni, Temeke and Illala. Street vendors in Dar es Salaam are governed by different bylaws in each of the city's three municipali-ties, but all the municipalities restrict street vending and make it almost impos-sible to obtain a licence or permit to trade legally.

Kinondoni Municipal Council bylaws

Kinondoni Municipal Council (KMC) has made a number of bylaws that affect street vendors, regulating access to markets and prohibiting businesses in areas other than designated markets.

KMC (Fees and Charges) Bylaws, 2004, allow the council to regulate fees and charges for services, licences and permits, at specified rates. Licences and permits are required for activities such as: ice-cream parlours; electronic gadget repairs; shoe making; foodstuffs vending; street markets, informal dry cleaners; tailors; oil selling in kiosks; auctions; informal carpentry; street barbers; second-hand clothing vending and non-specified businesses. Anyone vending on the street without a licence or permit contravenes the law. If a suspect admits an offence, the Municipal Executive Director can impose a fine, and should give an official receipt. Fines should not exceed the level imposed had the case gone to court. Courts usually impose fines of up to TSh 50,000 and/or imprisonment of 12 months.

KMC (Environmental Management) Bylaws, 2002, deal with environmental and health matters. Under the bylaws, street vendors trading in food, car washing or repairs must get permission from KMC before starting a business. However, street vendors have difficulty in obtaining a licence, due to lack of funds for the application fees and licence fees.

KMC (Waste Management and Refuse Collection Fees) Bylaws, 2000, deal with management of solid and liquid waste, refuse deposit and hazardous waste etc. Rule 11(2) restricts the use of packaging likely to cause waste, including using grasses, tree leaves and plant material for packing perishable goods, despite the fact that these materials are easily affordable for food vendors. The bylaws also prohibit nuisance on or near any premise or dwelling. Occupiers can be served notice to remove a nuisance caused by trade or domestic refuse. Thus, KMC may prohibit street vending using this bylaw on the pretext that it is nuisance to other street users and that it affects the cleanness of the street. Offences under these bylaws are enforced by either fines or custodial sentence, and any person convicted is be liable to a fine of up to TSh 50,000 or imprisonment of 12 months.

Temeke Municipal Council bylaws

Temeke Municipal (Markets Levies) ByLaws, GN 309 of 2010: Temeke Municipality Council (TMC) has introduced bylaws which empower TMC to establish markets and requires all market users to have a lease agreement. Rule 11(1) prohibits street vending without a permit outside a designated market, making any vending without a permit or outside a market illegal, which leaves street vendors at risk of harassment, arrests, confiscation of goods and having to bribe the law enforcers all the time. Penalties include a fine of up to TSh 50,000 or imprisonment of up to six months.

Temeke Municipal (Environmental Pollution Control) Bylaws, GN 310 of 2010 are designed to protect the environment in Temeke, and prohibit business operations, specifically selling food, without written permission from TMC. Often street vendors do not have enough capital to afford a permit, again leaving them vulnerable to harassment or bribes. Similar penalties can be imposed.

Temeke Municipal (Road Usage/Traffic Control) Bylaws, GN 311 of 2010 are intended ensure proper management and use of roads, specifying that '[i]t shall be prohibited for any person to conduct business activities in the road reserves or other areas not designated for that purpose without the permission of the municipal council'. Provisions of these regulations directly affect the rights of street vendors as there is no way they can do business legally. TMC also prohibits any car washing or vehicle repairs along the road, in road reserves or undesignated areas. For car-related violations, the bylaws impose penalties both on the person doing the activity and the vehicle owner.

Ilala Municipal Council bylaws

IMC (Environmental Cleanliness) Bylaws GN No. 111 of 2011: Ilala Municipal Council (IMC) covers the city centre and surrounding areas. These bylaws require any person who wants to do business in Ilala to have a permit or licence from IMC, and the bylaws prohibit anyone from operating any business in areas not specifically planned for that business, but there is no mechanism for establishing where businesses should be allowed, which affects vendors who need access to customers. Food-sellers need a permit before starting operations, which can be revoked if conditions are infringed. Similar penalties of fines of up to TSh 50,000 or 12 months imprisonment can be imposed.

IMC (Road Use) Bylaws, GN No.108 of 2011 provide for use of the roads in the IMC area and prohibit vending adjacent to the road, in order to prevent disruption to passage along the road. The bylaws also prohibit closure of the road without a written permission by IMC. Thus, blocking free passage by street vendors' merchandise constitutes an offence under the bylaws. Car-wash activities along the road or in open space are completely outlawed and, again, both the business operator and vehicle owner can be penalised.

IMC (Market Levies) Bylaws, GN No.107 of 2011 regulate all market businesses in IMC. All business people operating in the market must be registered by the market officer assisted by the committee of market stakeholders. The bylaws also restrict any unregistered businessman from doing business in the market except with permission of the market officer. Notably, the bylaws also outlaw any business done outside designated areas, which means street vendors – who trade on the roadside, open areas and other public spaces – are trading illegally.

Conclusion

Despite its long tradition, street vending in Tanzania since the late 1960s has been subject to ambivalent national policy that has wavered from tolerance and accommodation to crackdown and repression. The development of mass informal markets in urban Tanzania had its beginnings in the country's economic hardships of the 1970s and 1980s, and the government response of the 1980s, including the 1983 Penal Code and *nguvukazi* act, intended to restrict street trading and other informal work, was subverted to become enabling legislation as the new licensing system for petty trade and other activities provided a valuable source of municipal revenue. Since the late-1990s, there have been swings of government policy, moving from guarded recognition, for example following Hernando's de Soto's recommendations on formalisation, to the clearances of 2006–2007 and on-going removal of street traders for major development projects.

Many of the challenges faced by street vendors emanate from punitive laws that make compliance virtually impossible, particularly their access to land for trading. Using the example of Dar es Salaam as representing cities in Tanzania, it is evident that municipal bylaws in all the three municipalities of Dar es Salaam City, Kinondoni, Temeke and Ilala, inhibit street vending by imposing unattainable requirements for permits and licences; prohibitions against conducting business activities in areas not designated for that purpose and proscribing interferences with passage rights of vehicles and persons on the roadside and streets within their respective areas of jurisdiction.

The limitations which street vendors face in Tanzania may be compounded by the fact that Tanzania does not have a single unified legal framework that directly addresses the issue of street vendors, as evidenced by three municipalities in Dar es Salaam, each of the 132 districts in the country having a different set of laws affecting street vending. This also explains why there is no national street vending policy, which causes differentiated treatment of street vendors from one place to another. Thus, lack of policy and the impact of the existing legal framework on street vendors defy the provisions of the Constitution which protect the right to work, the right own property and guarantee the protection of such property.

Another challenge that the bylaws impose on the street vendors is that all have taken a penal approach in which street vending is considered illegal. They regard such activities as nuisance and against the orderly management of the munici-palities. This affects street vendors, whose conviction of an offence amounts to a criminal record, thereby limiting their credibility to obtain credit in the future.

Further, the law shows that the right to occupation and access to land is more theoretical than practical. While the Land Act advocates for the importance of giving right to access land to all citizens, municipal regulations provide to the contrary; access to land is inhibited in that permits and licences are required for any person to use land and street vendors' inability to acquire the permits and licences limits their access to land.

There is no legal provision that articulates how street vendors can access justice claiming their right to access land. The only avenue is through constitu-tional petitions where street vendors can claim their rights. The laws do not give space rights to street vendors. For instance, the Highway Act does prohibit doing business on the road reserves or raising any structures on the sidewalk. No excep-tion is provided for street vendors. The laws do not require urban authorities to accommodate street vendors in their plans. As a result street vendors find them-selves in informal areas and their businesses end up being considered illegal.

Therefore, the municipal law does not accommodate the global phenomenon and critical economic role that street vendors play. In cities and towns throughout the world millions of people earn their living by selling a wide range of goods and services on streets. Street vendors represent a significant share of the urban informal sector, but aversive regulation does not support their poverty-reduction potential in Tanzania.

Acknowledgements

Sections of the paper on Tanzania's legal framework draw on Ackson, T. (2014/15) and are reproduced with kind permission of the *SADC Law Journal*.

Notes

1　In 2003 the licence had 44 business/service entries in three groups (Tume ya Jiji, 1997). The first group had 16 entries and it includes vegetables, street foods, soft drinks, tea/coffee, ice-cream, fish and other perishable items.
2　Art. 7(2) of the Constitution explicitly provides that 'the provisions of this Part of this Chapter are not enforceable by any court. No court shall be competent to determine the question whether or not any action or omission by any person or any court, or any law or judgment complies with the provisions of this Part of this Chapter.'
3　See section 3(c) of Act No. 6 of 2007.

References

Ackson, T. (2014/15) Legal hostility towards street vendors in Tanzania: a constitutional quandary, *SADC Law Journal*, 4(1): 144–64.
Bagachwa, M. (ed.) (1994) *Poverty Alleviation in Tanzania: Research Issues*, Dar es Salaam: Dar es Salaam University Press.

BBC (2006) Tanzania suspends street clean-up, http://news.bbc.co.uk/1/hi/world/africa/4794332.stm, accessed January 2015.

Bernabe, S. (2002) Informal employment in countries in transition: a conceptual framework, CASE paper 56, Centre for Analysis of Social Exclusion, London School of Economics, http://euroscience.org/WGROUPS/YSC/BISCHENBERprep-session4c.pdf, accessed October 2012.

Brown, A. (2006) Setting the context: the social, political and economic influences on the informal sector in Ghana, Lesotho, Nepal and Tanzania, in Brown, A. (ed.), *Contested Space: Street Trading, Public Space and Livelihoods in Developing Cities*, Rugby: ITDG.

Burton, A. (2007) The haven of peace purged: tackling the undesirable and undesirable poor in Dar es Salaam, ca 1950–1980s, *International Journal of African Historical Studies*, 40(1): 119–51.

Campbell, H. and Stein, H. (1992) *Tanzania and the IMF: The Dynamics of Liberalization*, Boulder, CO: Westview Press.

Dill, B. (2007) Democracy, development and the paradox of associational life in Dar es Salaam, thesis, University of Minnesota.

ESRF (2006) Trade, development and poverty in Tanzania: when does trade reduce poverty, when doesn't it and why? Key messages on trade and poverty linkages in Tanzania, Economic and Social Research Foundation (ESRF) and Consumer Unit Trust Society (CUTS), http://www.cuts-citee.org/tdp/pdf/campaign_kits-tanzania.pdf, accessed October 2012.

Felister, P. (2012) Dar vendors, MP give TANESCO, DCC ultimatum, *Guardian*, 5 April, http://www.ippmedia.com/frontend/index.php/oi.poc/ilablten010-135546/?l=40207, accessed August 2015 [site no longer live].

Gibbon, P. (1993) *Social Change and Economic Reform in Africa*, Uppsala, Sweden: Scandinavian Institute of African Studies.

Horne, D.L. (1987) Passing the baton: presidential legacy of Julius K. Nyerere, *Journal of African Studies*, 14(3): 89–94.

ILO (2002) *Women and Men in the Informal Economy: A Statistical Picture*, ILO (International Labour Organization), http://wiego.org/publications/women-and-men-informal-economy-statistical-picture, accessed January 2015.

Jecha, T.K. and Mdundo, M.H.O. (1999) *Masimulizi ya Sheikh Thabit Kombo Jecha*, Dar es Salaam: DUP (1996).

Kahama, G., Maliyamkono, T.L. and Wells, S. (1986) *The Challenge for Tanzania's Economy*, London: James Currey; Dar es Salaam: Tanzania.

Kulaba, S.M. (1982) Rural settlement policies in Tanzania, *Habitat International*, 6(1–2): 15–25.

Lugalla, J.L.P. (1997) Development, change and poverty in the informal sector during the era of structural adjustments in Tanzania, *Canadian Journal of African Studies*, 31(3): 424–51.

Lyons, M. and Msoka, C. (2008) Micro-trading in urban mainland Tanzania: Final report – the way forward, unpublished research report, submitted to HTSPE Development Consulting Services.

Lyons, M. and Msoka, C. (2010) The World Bank and the street: (how) do 'Doing Business' reforms affect Tanzania's micro-traders?, *Urban Studies*, 47(5): 1079–97.

Maliyamkono, T.L. and Bagachwa, M.S.D. (1990) *The Second Economy in Tanzania*, London: James Currey.

Mbilinyi, S. (1974) Economic struggles of TANU government, in Ruhumbika, G. (1974), *Towards Ujamaa: Twenty Years of TANU Leadership*, Dar es Salaam: East African Literature Bureau.

Mhamba, R. and Titus, C. (2001) Reactions to deteriorating provision of public services in Dar es Salaam, in Tostensen, A., Tvedten, I. and Vaa, M. (2001) *Associational Life in African Cities: Popular Responses to the Urban Crisis*, Stockholm: Elanders Gotab.

Miti, K. (1987) *Whither Tanzania?*, Delhi: Ajanta.

MKURABITA (2008) The Property and Business Formalization Programme Reform Proposal Vol. 1: *Executive Summary*.

Msoka, C. (2005) Informal markets and urban development: a study of street vending in Dar es Salaam, Tanzania, thesis, Minneapolis: University of Minnesota, unpublished.

Mwananchi (2011) People to protest evictions in Ubungo, 6 March.

Ndanshau, M. and Mvungi, A. (2002) *The Congo Street Culture: Creation of Wealth and Employment through Petty Trade, Final Report*, Dar es Salaam: REPOA (Research on Poverty Alleviation).

Nnkya, T. (2006) An enabling framework? Governance and street trading in Dar es Salaam, Tanzania, in Brown, A. (ed.), *Contested Space: Street Trading, Public Space and Livelihoods in Developing Cities*, Rugby: ITDG.

Ntetema, V. (2006) New rules for Tanzanian traders, BBC News, http://news.bbc.co.uk/1/hi/world/africa/6059004.stm, accessed January 2015.

Nyerere, J.K. (1977) *The Arusha Declaration Ten Year Years After*, Dar es Salaam: Government Printer.

Nyerere, J.K. (1982) On rural development, *Habitat International*, 6(1–2): 2–24.

Osafo-Kwaako, P. (2011) Long-run effects of villagisation in Tanzania, http://www.econ.yale.edu/conference/neudc11/papers/paper_336.pdf, accessed September 2015.

Owens, G.R. (2010) Post-colonial migration: virtual culture, urban farming and new peri-urban growth in Dar es Salaam, Tanzania, 1975–2000, *Africa*, 80(2): 249–74.

Peter, C.M. (1997) *Human Rights in Tanzania: Cases and Materials*, Koln: Rudiger Koppe Verlag.

Sabahi (2013) Tanzania: Iringa parliamentarian, 60 traders arrested, allAfrica, http://allafrica.com/stories/201305210297.html, accessed January 2015.

Sanders, T. (2001) Save our skin: structural adjustment morality and occult in Tanzania, in Moore, H.L. and Sanders, T. (2001) *Magical Interpretations, Material Realities: Modernity Witchcraft and the Occult in Post-Colonial Africa*, London: Routledge.

Sayaka, O. (2006) 'Earning among friends': business practices and creed among petty traders in Tanzania, *African Studies Quarterly*, 9(1–2): 23–38.

Tanzania Daima (2012) More news emerges amid the death of a street vendor, Saturday, 17 November.

Tegambwage, N. (1985) *Duka ia kaya*, Dar es Salaam: Tausi.

TDN (2014) Tanzania: Machinga rioters wreak havoc in Mwanza, *TDN* (*Tanzania Daily News*), http://allafrica.com/stories/201406120183.html, accessed January 2015.

Tripp, A.M. (1996) Contesting the right to subsist: the urban informal economy in Tanzania, in Swantz, M.L. and Tripp, A.M., *What Went Right in Tanzania: People's Response to Directed Development*, Dar es Salaam: Dar es Salaam University Press.

Tripp, A.M. (1997) *Changing the Rules: The Politics of Liberalization and Urban Informal Economy in Tanzania*, Berkeley: University of California Press.

Tume ya Jiji (1997) Mwongozo kwa wafanya biashara ndogondogo (Guide for small busi-nesses), Dar es salaam: Tume ya Jiji la Dar es Salaam (Dar Es Salaam City Commission) Mei, unpublished.

URT (1977) United Republic of Tanzania Constitution of 1977.

URT (1999) Land Act, Cap 113, R.E. 2002.

URT (2007a) Land Use Planning Act, No. 6 of 2007.

URT (2007b) Urban Planning Act, No. 8 of 2007.

Wagao, J.H. (1990) *Adjustment Policies in Tanzania, 1981–1989: The Impact on Growth, Structure and Human Welfare*, Florence: International Child Development Centre.

WIEGO (2016) Statistical picture, http://wiego.org/informal-economy/statistical-picture, accessed April 2016.

WB (World Bank) (2002) Upgrading of low income settlements, country assessment report, Tanzania, http://web.mit.edu/urbanupgrading/upgrading/case-examples/overview-africa/country-assessments/reports/Tanzania-report.html, accessed August 2015.

WB (2004) Projects and operations: urban sector rehabilitation project, http://www.worldbank.org/projects/P002758/urban-sector-rehabilitation-project?lang=en, accessed August 2015.

Young, R. (1983) *Canadian Development Assistance to Tanzania*, Ottawa: North–South Institute.

13 Street trade in post-Arab Spring Tunisia

Transition and the law

Annali Kristiansen, Alison Brown and Fatma Raâch

Summary

Tunisia's 'uprising for dignity and democracy' of 2011 was ignited by the death of street vendor Mohamed Bouazizi in January 2011. This led to mixed fortunes for street vendors. This chapter explores the socio-economic context in which the uprising took place, the legal framework for *commerçants détaillant ambulants* in Tunisia, and how street vendors perceive the effects of the law on their trade. The chapter concludes that, despite a long history of partial acceptance of the role of the informal economy in Tunisia, experience on the streets remains fragile.

Introduction

Tunisia is going through a period of transition. In 2011, in the Jasmine Revolution mass demonstrations took place in response to the self-immolation of Mohamed Bouazizi, a street vendor working on the brink of legality in Sidi Bouzid in Tunisia's interior, protesting at constant harassment and humiliation from police and city officials that denied him the means to survive. His death sparked nation-wide protests that unseated President Ben Ali, who held power for 24 years, and inspired revolution throughout the Arab world. Alone in the region, Tunisia's revolution has led to peaceful political and democratic change with an elected parliament and president, and a coalition government under a new constitution (*Guardian*, 2014).

The present study was undertaken during a period when the constitution of the country was debated and the entire legal system was under review. Such a context affects how law and regulation are developed and applied. Nevertheless, the chapter examines the legal framework within which street trading operated, and the challenges to street vendors that it posed.

The chapter is set out in three parts. The first examines the socio-economic and employment context in Tunisia, to explore the background to the revolution and the context of life in Tunisia. The second section analyses the legal framework and texts that regulate street vending, guided by informal exchanges with municipal police officers who enforce regulations on the occupation of public space. The third section provides a brief analysis of the experiences of street vendors with the law from a study in Tunis in 2013. The study was carried out under a 2012 British Academy small grant project, entitled *Economic Inclusion and Political Change: Impact on Street Trading of Regime Change in Tunis and Cairo.*

General context

The unfavourable and precarious conditions of street vending constituted, and continue to constitute, an essential vindication of the voices of the poor, who call for security and recognition of their work. This resonates particularly well with the revolution's objectives of work, liberty and dignity. It also shows the relation between the abstract economic, social and cultural rights, as set out in the *International Covenant on Economic, Social and Cultural Rights*, and the issue of realising a decent life and dignified living conditions in Tunisia.

The 2011 uprising created the possibility of a new Tunisia under a new rule. Since the departure of President Ben Ali, the country has witnessed the establishment of successive interim governments, with a 'small constitution' and existing legal texts providing the main legal framework. After the elections in 2011 the troika[1] came into power and the National Constituent Assembly was established. By the summer of 2013 it became clear that additional efforts were required to ensure that the aims of the uprising were met and that the foundations for security created.

Civil society actors played a key part in this transformation. The Tunisian General Labour Union (UGTT), the Tunisian Confederation of Industry, Trade and Handicrafts (UTICA), the Tunisian League of Human Rights (LTDH) and the National Organisation of Tunisian Lawyers (ONAT)[2] joined forces in a quartet that commenced the national dialogue to establish a common response to the Tunisian challenge. By the end of 2013 a roadmap was agreed with the government and Constituent Assembly, resulting in the adoption of the Constitution of Tunisia and the establishment of a technocrat government at the end of January 2014. After three years of uncertainty and debate about governance, the new Constitution of 2014 was passed, which largely builds on the rule of law and human rights, whereby the state ensures the conditions for a decent life (RT, 2014: Art. 21). In 2015, the Nobel Peace Prize was awarded to the quartet for their role in ensuring a peaceful transition to democratically elected government.

Employment and unemployment

A weak economy and poor employment prospects have been at the heart of the unrest. Although average for the Middle East and North Africa, Tunisia's employment ratio (41% in 2013) is low, mainly because few women are employed or participate in the labour market, and because many young people remain in education due to the difficulty of finding a job (LO/FTF Council, 2015: 7).

However, unemployment is very high, estimated at 16% in 2013. Youth unemployment is even higher at 36%, which means, effectively, that it is almost impossible for young people to find jobs. There are also large regional differences in unemployment, with the lowest being in the central east region (11%) and the highest in the central-west (29%). About 39% of the unemployed have been unemployed for more than a year. It is alleged that areas bordering Libya have an unemployment rate of almost 50% (ibid.: 8–9).

The state runs most social security systems in Tunisia, but those working in the informal economy are largely excluded. There are two contributory social security schemes, the *Caisse Nationale de Sécurité Sociale* (CNSS) for the private sector and the *Caisse Nationale de Retraite et de Prévoyance Sociale* (CNRPS) for the public sector. According to 2010 data, 27% of the population had no schooling (ibid.: 3, 13).

Informal trade and street vending

For many years, the role of the informal sector in Tunisia has been recognised, but, since 2011, it has become apparent that many types of trade exist and continue to develop. New forms of street vending have emerged, such as the provision of repair services. The high unemployment means that street vending appears to be more popular than ever.

There are few reliable estimates of the scale or economic contribution of the informal economy, but analysis suggests that it may include around half a million workers (15% of the employed) – labour laws do not cover these workers. Of own-account workers, 2010/11 estimates suggested that 52% were in services and trade (mainly street vendors) (INS, 2010/11). A study undertaken through the Tunisian Inclusive Labour Initiative (TILI) interviewed 1,203 workers across six regions, including own-account workers, employers and employees. Some 90% of employers and own-account workers did not have a licence; for these bureaucracy, corruption and the cost taxes were major obstacles to formalisation (GFI *et al.*, 2013). By 2013, a study by the Tunisian Association for Management and Social Sustainability (TAMSS) found much higher levels of informal employment – 53% across Tunisia. Often informal workers are men with no school certificate aged between 25 and 54 years (TAMSS, 2013).

Figure 13.1 Commerce de la débrouille in Tunis

Street vending covers various types of trade Tunisia, and is generally known as the *commerce de la débrouille* or 'the trade of coping and resourcefulness' (International Crisis Group, 2013). In addition to the tradition of selling home-made produce, there are thousands of destitute young people, often from poorer regions, who become 'petty traders' (*petits marchands*) or 'vendors on the sly' (*vendeurs a la sauvette*) who offer retail products at makeshift stalls in the urban centres of Tunisia.

Abderrahmen Ben Zakkour, Professor of Economics, University of Tunis, has undertaken numerous interviews with those in precarious work. His study of 264 women working in the informal economy revealed the prevalence of job insecurity and poverty, and the continuous struggle for subsistence. Beyond the need for sustenance, he found three critical financial concerns: accommodation costs (rent or payments to the owner), the cost of schooling for children and cost of health care due to the lack of adequate social security cover (Ben Zakkour, 2010–11). He identifies the following categories of street vendors:

1) *micro-entreprises sur la rue*: it is estimated that there are more than 500,000 micro-enterprises trading from the street in Tunisia;
2) *commerce ambulante*: people with a small cart (the municipality often gives them an authorisation but does not keep a record of the number of permits);
3) *marchés hebdomadaires*: weekly markets provide an outlet for food producers; most are quite well organised and vendors rent space;
4) *femmes de la maison*: home-based workers make food, clothing or household goods, which are sold at the weekly markets – 90% of these workers are women;
5) *service de réparation*: repair services provided from a small truck; this is a new category, which has arisen since the revolution (interview, 2013);
6) *économie frontalier*: the border economies between Tunisia/Algeria/Libya. Goods commonly exchanged include petrol from Algeria to Tunisia; and wheat and grain from Tunisia to Algeria.

Street vending is generally regulated by law, and the categories of 1) micro-enterprises on the street, 2) street vendors and 4) weekly markets are broadly recognised. However, in the aftermath of the uprising another type of business has grown, that of the *marché parallel* (parallel market), partly consisting of the *économie frontalier* – the smuggling which falls outside any legal framework. The *marché parallel* is said to have started more than 30 years ago, originally involving trade of tinned food and new types of food, such as mango and coconuts, products which are now available legally in Tunisia (La Presse de Tunisie, 2014: 18).

The International Crisis Group (2013) finds that:

> the majority of inhabitants in the border regions do not consider these small transporters and '*passeurs*' as smugglers (*knatri*). They prefer to use the

name 'traders' or 'entrepreneurs'. The term 'bootlegger' is reserved for the rather mysterious and mythical figure of the '*son of the frontiers*', who takes the risk of illegally crossing the borders to import highly-taxed products that are prohibited, such as cigarettes, alcohol and raw materials (petrol, construction iron, copper, etc.).

It is generally estimated that the *marché parallel* constitutes 7% of the Tunisian economy. According to the World Bank, informal trade accounts for 5% of all imports; however, informal trade represents an important part of Tunisia's bilateral trade with Libya and Algeria, accounting for more than half of the official trade with Libya and more than the total of the official trade with Algeria (Ayadi *et al.*, 2013).

A study by the Ministry of Trade indicates that 77.6% of Tunisians obtain products from the *marché parallel*, thus contributing 15–20% of the gross national product (GDP, or *Produit Intérieur Brut*, PIB) and employing more than 31% of manpower outside agricultural production. It is alleged that approximately 14,000 jobs exist in supply networks alone. According to the Ministry of Trade the *marché parallel* exists in 'illicit points of sale' (46%), in commerce (21%), in connection with storage (18%), living and similar spaces (15%) (Les chroniques des tunipage, 2014).

Street vending is thus a part of a large trade in Tunisia that appears to have gained in volume and importance since the 2011 uprising, and to be continuing to do so.

The local administration and planning

The local administration in Tunisia comprises the *collectivités locales*. Historically, the *cheikhats* were the local seat of power at the tribal level. Originally clan-based, the Ottoman Empire transformed them into a territorial local administration, with local police power in some places, which over time undermined the independence and legitimacy of the *cheikhats*.

Drawing on French civil law, the protectorate introduced *communes*[3] at town and city level, and introduced cultural heritage protection. Urban planning commenced with the Decree of 25 January 1929 that aimed to satisfy collective hygiene, free movement and aesthetics (Mrad, 2014). In the 1940s, three decrees were introduced that addressed local power and planning regarding: planning and architecture; requirement for authorisation to build; and rural planning. Together they formed a legal framework for rural and urban communes in regard to planning and authorisation (ibid.).[4]

In terms of local administration and decision-making, *collectivités locales* have over time been granted powers over licensing, public service provision and taxation (Snoussi, 1958), and, since 1956, urban regulation has developed into a complex of administrative law and planning. Law 94-122 of 28 November 1994, on the development of the territory and urbanism, provided a vision of urban planning and remains a central urban law in Tunisia.

A Municipal Charter on local administration came into effect in 1952 and, with the independence of Tunisia in 1956, *collectivités locales* were established at the level of municipalities or communes. The municipal law, which came into effect on 14 March 1957, aimed to promote decentralisation and created an essentially Tunisian system of municipal councils and assemblies. Shortly after, legal reform debates commenced that sought to soften the hierarchy of the administration and legislation, and to establish which agency determined fees and taxes – for example, for the occupation of public space (ibid.). Today, the Constitution of 2014 (RT, 2014: Chapter VII) provides for decentralised local power, exercised through the *collectivités locales* (municipalities, regions and departments). The *collectivités locales* have a legal personality, financial and administrative autonomy, and manage local affairs based on the principle of freedom in administration. Both municipal and regional councils are elected by universal suffrage, and they in turn elect the members of district councils.

The competence of the *collectivités locales* can be exercised jointly with the central authority, which can apply the principle of subsidiarity and transfer competences and resources to the *collectivités locales*. In addition to local management of resources, the *collectivités locales* should comply with the principles of good governance (including financial control) and adopt tools for democratic participation and open governance to involve citizens and civil society in planning, implementation and monitoring of development projects.

Although street vendors are most likely to come into contact with *collectivités locales* over authorisation, regulation and enforcement, Tunisia's transition and socio-economic context suggests that informal trade and street vending could further develop.

The legal framework of street vending

This section outlines the main legal framework affecting street vending in Tunisia. The main legal texts that organise or address the exercise of street vending are as follows. Names are translated from the French, and full titles, as found in the *Journal officiel de la république tunisienne* (*JORT*), are given at the end of the chapter.

- *Law No. 2009-69 of 12 August 2009* relating to retail commerce, JORT 14 August 2009, No. 65 (JORT, 2009).
- *Law No. 11-1997 of 3 February 1997* on the promulgation of the law of local taxation (JORT, 1997).
- *Law No. 74-33 of 14 May 1975* on the promulgation of the organic law of communes.
- *Decree No. 203236 of 2 August 2013* on the modification of Decree No. 98-1428 of 13 July 1998 on the establishment of taxes that the *collectivités locales* are authorised to collect.
- *Decree by the Minister of Interior and Local Development and the Minister of Trade and Handicrafts of 9 December 2010* establishing the conditions

and the procedures of exercising street vending, JORT 14 December 2010, No. 100 (JORT, 2010a).

- *Decree by the Minister of Trade and Handicrafts of 9 December 2010*, completing the Decree by the Minister of Trade and Handicrafts of 18 June 2005 on the administrative services provided by the offices pertaining to the Ministry of Trade and Handicrafts and the public businesses and establishments under its administration (*sous tutelle*) and the conditions of granting authorisation, JORT 14 December 2010, No. 100 (JORT, 2010b).
- *Decree No. 98-1428 of 13 July 1998* in regard to the establishment of the taxes (tariffs) that the local collectives (*les collectivités locales*) are authorised to claim. JORT 24 July 1998, No. 59 (JORT, 1998). (There is more recent tax legislation, but it does not touch directly on street vending, e.g. Law No. 53-2007 of 8 August 2007.)

Definition and conditions of street vending

The legal framework outlines the *conditions of street vending*, the attainment of the title of street vendor and the regulation of the required authorisations.

Law No 2009-69 of 12 August 2009 on the distribution trade sets out the principle of free trade and regulates the exercise of the activities related to trade distribution. Art. 1 stipulates that: 'The law herein sets the rules governing the practice of distribution trade according to which freedom is the principle and authorisation is the exception' (JORT, 2009). Thus the principle of free trade exists, but commercial activity is, in effect, regulated in accordance with the legal framework.

The conditions for the exercise of street vending are set out in Art. 9 of Law No. 2009-69 which states that,

> Itinerant retailing may be practised after getting an itinerant retailer certificate. The terms and procedures of practising this activity shall be set by joint order of the Minister of the Interior and Local Development and the Minister in charge of trade (commerce).
>
> Ibid.

At that time, the Ministry of Interior did not cover local development, but the 2014 government transferred this to the Ministry of Equipment, Territorial Planning and Sustainable Development and the Ministry of Commerce and Handicrafts.

In addition, Art. 9 outlines the elements that *define street vending*, as:

1) A physical person who does not have a permanent space or premises in which to exercise the retail profession.
2) The retail of products that takes place in spaces intended for the purpose and using 'equipment that may be dismantled or transported'.

The term adopted in Law No. 2009-69 for a street vendor is '*commerçant détaillant ambulant*', literally translated as 'itinerant retailer'.

The definition of a street vendor is further developed by the Decree of 9 December 2010 (JORT, 2010). Art. 2 determines the conditions and the procedures for exercising street vending. This decree applies the same definition as Law No. 2009-69 and it specifies that, in order to carry out this profession, 'it is considered that "equipment that may be dismantled or transported" includes the use of stands or stalls or means of transport and towing carts/trailers that have been equipped for this purpose' (ibid.: Art. 2). This mobile retailing should take place in reserved spaces, including weekly markets and spaces fitted out for the purpose.

There are two conditions to be met in order to practise street vending (JORT, 2010):

1) The vendor must obtain a permit as a vendor.
2) The vendor must respect the regulations concerning the occupation of public space in particular as mobility is an essential aspect of this business.

Street vendor authorisation

A further question is how to acquire an *authorisation as a street vendor*. Art. 4 of Decree of 9 December 2010 foresees that 'any person who requests permission to exercise itinerant retailing, must obtain an itinerant retail permit or card as issued by the regional directorate of commerce, based upon the opinion of the governor of the territorial jurisdiction'.

Art. 7 states that 'Any person who applies for an itinerant retail permit must submit a file to the regional directorate of commerce with the territorial jurisdiction.' It thus confirms that an authorisation is a condition for exercising street vending. Only persons of 18 years or more, with a residence permit and a national identity card, can apply.

In addition, an 'authorisation of occupation of the public domains' (of roads, the governorate, the municipality, or the public maritime domain, issued by the competent authority) is sometimes required, and a copy must then be submitted by the applicant (Art. 7). According to some, municipalities often give authorisations, but they do not keep a record of these, which could imply that it is only the holder of an authorisation that can prove that he/she is a legally entitled street vendor.

Several groups are exempt from the requirement of an authorisation or street vendor card. Art. 6 states that:

> the craftsman and agricultural producers who would like to sell their own produce directly at consumers' weekly markets are exempt from the requirement of obtaining an authorisation or a card on the condition that they can prove the quality of the produce to the entity in charge of managing the market.

The sale of local or homemade produce is very common, and includes bread baked and sold by women and homemade honey and breadsticks, which are often sold by men. This produce is regularly sold at the roadside, near markets, or outside supermarkets.

Thus, several categories of informal trade and street vending are regulated by the law, including: micro-enterprises on the street; street vendors; and weekly and other markets. However, recent years have seen a development of other types of activity, such as repair services in the street, which are not a retail activity and therefore not necessarily covered by the law.

Authorisation to occupy public space

The legal *occupation of public space* is a condition for the authorised exercise of street vending. Law No. 75-33 of 14 May 1975 (JORT, 1975) on the promulgation of the organic law of communes, confers to municipalities the competence on the classification and regulation of urban roads and by-ways.

Art. 107 of Law No. 75-33 stipulates that:

> the roads and by-ways (*la voirie vicinale*) incorporated into the communal domain are classified as public urban roads and by-ways. Public urban roads provide service within the interior of the agglomeration and the by-ways provide the link between the agglomerations. The classification of public urban roads or by-ways is a result of the municipal development plan, which has been legally approved; or failing that by decree made by the Chairman of the Municipal Council that is made after a Council deliberation and based upon the opinion of the services of the Ministry in charge of urbanism.

Taxation or fees for the occupation of space

The law also confers on *collectivités locales* the power to claim taxes for the occupation of markets that are situated within their jurisdiction, as outlined in Decree No. 98-1428, related to the establishment of tax tariffs by *collectivités locales* (JORT, 1998)[5] and Law No. 11-1997 on the promulgation of the code of local taxation (JORT, 1977).

Law No. 11-1997 addresses many aspects of local taxation, and most tariffs for fees are established by decree, except for those for 'collective stationing' (JORT, 2013). Chapter VIII, Section 3 of Law No. 11-1997 concerns the rights in markets – relating to 'collective stationing'. There is a general right to 'station' goods, animals and commodities of all kind for sale at daily, weekly, occasional or wholesale markets, or in spaces delimited by the *collectivité locale* for the meeting of sellers and buyers. This right is at the expense of the seller.

For daily, weekly or occasional markets, *collectivités locales* can establish a particular 'right of stationing' and fix the fee by decree. The authority may approve tutelage (management fees) if the general 'right of stationing' results in disproportionate costs in managing the market (Art. 69). The traditional trade in homemade produce is exempt from this tax (Art. 71).

For street vending, the fee is also payable to *collectivités locales* for temporary occupation of public thoroughfares by: coffee stalls, restaurateurs, display works, all persons that exercise a trade within an installation that is mobile or can be dismantled and vehicles parked for the transportation of persons or merchandise. The fee shall be paid within the same time limit and according to the same criteria as for the occupation of public space (Art. 85).

In addition, the law outlines the fees for public services that are provided by the *collectivités locales*, such as the maintenance of sanitation systems, street lighting and waste collection from businesses and industries, etc. (Art. 91).

From the above it may appear that both informal trade and street vending, as well as other activities, such as repair services, could be defined as 'the exercise of a trade within the framework of an installation that is mobile or can be disassembled'. It thus appears as if taxation by *collectivités locales* for temporary occupation of public space covers a wide span of activities.

Breach of conditions

If the legal conditions for street vending are not respected, then there are legal sanctions. Under Art. 32 of Law No. 2009-69, anyone working as a street vendor without the required permit or card as provided for in Art. 9 of Law No. 2009-69 is liable for 'a fine of between 500 and 3,000 (Tunisian) Dinars and the seizure of the products'.

At the same time, there is some flexibility in the case of a possible infraction. Art. 35 of Law No. 2009-69 allows the person charged with breaching the law to 'request settlement with the Ministry in charge of commerce provided that a judgment has not been pronounced'. Along the same lines, Art. 13 of Decree of 9 December 2010 also refers to Art. 32 on the regulation of cases that are potentially in breach of the law (JORT, 2010a).

It appears that there is some flexibility given for potential authors of infractions to regularise their situation. While this possibility exists, the actual number of regularisations is, in fact, very limited, mainly as it is so difficult to obtain an authorisation for the occupation of public space. While there are sanctions to be applied in the case of breach of procedure, law enforcement officers indicate that they try to address infractions in a sympathetic manner, to avoid conflict with the street vendors. This should be seen in a context of socio-economic challenges and at a time where (formal) employment is scarce (key informant interviews).

The legal framework as seen by street vendors

As part of this research, surveys were conducted with traders in Tunis to examine their context of trading and the impact of the Arab revolution. The survey took place in three main localities in early 2013: around the Medina including Rue d'Espagne, Bab Aljazira and Ibn Khaldoun; Marché Sidi Bahri, near the Ali Balhouene bus station; and the low-income suburban area of Ettadhamen and

Intilaka. Interviews were undertaken with 103 traders, 88 men and 15 women. Some 34% of interviewees were aged 19–34 and 35% aged 35–54. Many interviewees were well-educated: 41% had secondary education, five had a diploma and one a university degree, while 13% had no schooling. Some 83% of interviewees were the main income earners in their households.

The survey explored relations with municipal authorities and police, which often define how traders view their security. A number of dimensions of insecurity were explored, including harassment (usually by the authorities), confiscation of goods, fines, evictions, prosecutions and relocation. Before the revolution, nearly half (48%) of all interviewees had had goods confiscated, over a third (35%) had suffered eviction and 12% had been prosecuted. After the revolution there was a marked drop, although the problems remained significant: confiscations fell to 17%, evictions to 12% and prosecutions to 9% afterwards.

Many traders were resigned to the illegality of their occupation.

> *J'ai été arrêté plus d'une fois, et eu la confiscation des marchandises. Je les ai repris à payer à la municipalité. Nous avons été exposés à la violence verbale et physique par la police municipale. Les problèmes abondent – le harcèlement et le chaos ont provoqué la diminution du revenu et la récession du marché.*
>
> Man, selling jewellery
>
> (I was arrested more than once, had confiscation of goods. I took them back by paying the municipality. We've suffered exposure to verbal and physical abuse by the municipal police. Problems abound, and the harassment and chaos has caused a drop in income and market recession.)

Some had managed a degree of accommodation with the police:

> *C'était plus facile. Les vendeurs donnaient de l'argent aux agents de la municipalité pour libérer les marchandises confisqués. Soumis à la violence au cours des affrontements et des manifestations. Perturbation de notre travail au cours de la révolution, et en cas de grève.*
>
> Man, selling tablecloths
>
> It was easier (before the revolution). The vendors gave money to the municipal agents to obtain the confiscated goods. We suffered violence during the clashes and demonstrations. Our work was upset during the revolution and in the case of strikes.

However, others felt unprotected by the law:

> *Les problèmes sont nombreux: conflit continu entre les marchands et la police municipale, dans un état d'absence de tout droit qui garantit le droit de travail pour le marchand.*
>
> Man, selling Chinese-made TV remote controls

The problems are numerous: ongoing conflict between the merchants and the municipal police, in the absence of any law which guarantees the right to work for vendors.

Vendors faced many problems, including:

> *l'absence d'un cadre juridique qui protège ce travail; l'exposition à un grand nombre de difficultés financières la sécurité et le climat (précipitations et température); le manque à gagner; la concurrence et la récession; l'entrée du pays dans des grèves et des manifestations entraîne la violence et l'arrêt du travail pendant plusieurs jours et aggrave la situation.*

Man, selling children's magazines
and household goods

the absence of a legal framework that protects this work; exposure to a large number of financial difficulties; security and the weather (rain and temperature); the lack of earnings; competition and recession. The strikes and demonstrations in the country have led to violence, which meant stopping work for several days and made matters worse.

Conclusions

The uprising of Tunisia and the upheaval in addressing governance, rights and responsibilities has had an effect on life in general. The transition has meant that future life and livelihoods are being questioned by many Tunisians due to the lack of socio-economic security, rapid rise in unemployment and link between services such as social security and formal employment.

Tunisia has long recognised the prevalence of its informal economy, and government authorities have a reasonable recognition of the scale, economic potential and challenges of street vending. The *Institut National de la Statistique* (Institute of National Statistics, INS) undertakes regular employment surveys, which include analysis of the informal sector, and there have been several authoritative studies of the *marché parallel* with a clear distinction between that and more benign activities in the informal economy. In addition, Tunisia has strong trade unions, which show an increasing recognition of the need to extend social protection to vulnerable workers.

Informal work appears to be increasing in scale and importance since the uprising in 2011, and street vending is part of a wider trend in Tunisia. Thus street vending is fast becoming integrated into society, and *la commerce de la débrouille* (trade of resourcefulness) describes clearly that many street vendors 'just get by'.

The legal framework for street vending is simple and complex at the same time. While applying for a street vendor authorisation *per se* seems simple and clear, it is the link with the application for the authorisation to occupy public space that is a challenge, and it is generally up to the Chairman of the Municipal Council to authorise the occupation of public space. The question is how these authorisations are, in fact, conferred, particularly as the authorisation to occupy public space is usually a precondition for applying for a street vending licence.

In part, the legal framework provides an enabling context for the operation of street vending. The *commerçants détaillant ambulants* (itinerant retailers) include micro-enterprises on the street, street vendors and traders at weekly markets. These are regulated by the law and can obtain licences from the *collectivités locales* to trade in designated markets or public space. Other activities, such as repair services, do not necessarily fall within the legal framework of vending authorisation, but can be authorised and subject to local tax if approved by the *collectivité locale.*

There is also flexibility in both the legislation and its operation. While the law allows for fines and sanctions if the authorisation regime is flouted, there is some scope for accommodation if those accused offer to make a settlement. Law enforcement officers also indicate that they try to address infractions in a sympathetic manner and to be lenient in order not to enter into conflict with street vendors. Nevertheless, for street vendors, their experience is mixed. Police harassment declined noticeably in the aftermath of the revolution, but many felt their protection under the law remained weak.

The 2014 Constitution aims to decentralise power to the local level, and the constitutional emphasis on *collectivités locales,* their composition and competence could have a positive effect on local administration and decision-making – and thus potentially on local street vending, although street vendors are most likely to come into contact with *collectivités locales* over issues such as licensing, regulation or enforcement. Tunisia's socio-economic context and transition could imply that the informal economy and street vending will develop further. The way in which the 2014 Constitution will take effect and, in particular, the outcome of the local elections in 2017 remains to be seen, especially at the local level, but clearly there is some way to go before adequate accommodation is made for street vendors as a vulnerable sector of the population.

Notes

1   The 'troika' was the unofficial name for the alliance between the three parties that ruled Tunisia in the interim after the 2011 Constituent Assembly election.
2   *Union Générale Tunisienne du Travail* (UGTT); *Union Tunisienne de l'Industrie, du Commerce et de l'Artisanat* (UTICA) ; *Ligue tunisienne des droits de l'homme* (LTDH); *Ordre National des Avocats de Tunisie* (ONAT).
3   A *commune* is a *collectivité locale* that manages municipal interests (JORT: 2006, Art. 1) within a specific area (JORT: 2006, Art. 2 and chapter II).
4   Decrees of 10 September 1943, 22 July 1943 and 11 January 1945 (Mrad, 2014)
5   JORT n° 59 of 17 July 1998.

References

Ayadi, L., Benjamin, N., Bensassi, S. and Raballand, G. (2013) Estimating informal trade across Tunisia's land borders, Policy Research Working Paper 6731, World Bank,

Middle East and North Africa Region, http://documents.worldbank.org/curated/en/2013/12/18702424/estimating-informal-trade-across-tunisias-land-borders, accessed April 2016.

Ben Zakkour, A. (2010–11) *Les femmes et le travail informel: dans le banlieues populaires du Grand Tunis*, Tunis: Association des femmes Tunisiennes pur la recherche et le developpement/Espace Tanassof.

Guardian (2014) The Guardian view on Tunisia's success story, 26 December, http://www.theguardian.com/commentisfree/2014/dec/26/guardian-view-tunisia-transition-success-story, accessed April 2016.

GFI, ITIS, TAMSS (2013) Survey of informal workers in Tunisia, Study report, GFI (Global Fairness Initiative), ITIS, TAMSS (Tunisian Association for Management and Social Sustainability), http://www.globalfairness.org/attachments/article/118/FINAL%20SURVEY%20REPORT%20ENGLISH.pdf, accessed March 2016.

INS (2010/11) *Tunisie, Enquête nationale sur l'emploi, 2010 and 2011*, Tunis: INS (Institut National de la Statistique).

International Crisis Group (2013) La Tunisie des frontières: jihad et contrebande, *Rapport Moyen-Orient/Afrique du Nord*, 148, 28 November, footnote 44.

La Presse de Tunisie (2014) Effluves épicés: les vertus de l'informel, Monday, 3 March: 18.

Les chroniques des tunipages (2014) Revue de veille strategique trimestrielle, 3, January–February–March: 4–5.

LO/FTF Council (2015) Tunisia – Labour Market Profile 2015 (Danish Trade Union Council for International Development Co-operation), http://www.ulandssekretariatet.dk/content/landeanalyser, accessed April 2015.

Mrad, A.A. (2014) *Droit de l'Urbanisme*, Tunis: Latrach Editions.

Snoussi, M. (1958) Les collectivités locales en Tunisie, Mohamed Snoussi, Tunis, juillet 1958. UNESCO/SS/22, Paris, 14 November.

TAMSS (2013) Le phénomène de l'emploi informel – entre réalité et solutions, http://www.lexpertjournal.net/fr/385/societe-enquete-sur-le-phenomene-de-lemploi-informel-en-tunisie-entre-realites-et-solutions-778-des-employes-informels-gagnent-moins-de-200-dinars-par-mois-30-disent-quils-n/, accessed April 2016.

Legislation

Available from: http://www.legislation.tn/en/recherche/legislatifs-reglementaires

RT (2014) La Constitution de la République Tunisienne, Chapitre II: Droits et libertés, Art. 21.

JORT (2010a) Arrêté du ministre de l'intérieur et du développement local et du ministre du commerce et de l'artisanat du 9 décembre 2010, portant fixation des conditions et des procédures de l'exercice de l'activité du commerce de détail ambulant, JORT du 14 décembre 2010, n° 100.

JORT (2010b) Arrêté du ministre du commerce et de l'artisanat du 9 décembre 2010, complétant l'arrêté du ministre du commerce et de l'artisanat en date du 18 juin 2005 relatif aux prestations administratives rendues par les services relevant du ministère du commerce et de l'artisanat et les entreprises et établissements publics sous tutelle et aux conditions de leur octroi, JORT du 14 décembre 2010, n° 100, http://www.legislation.tn/sites/default/files/journal-officiel/2010/2010G/Jg1002010.pdf, accessed April 2016.

JORT (2009) La loi n° 2009-69 du 12 août 2009 relative au commerce de distribution, JORT 14 août 2009, n° 65, http://www.legislation.tn/en/detailtexte/Loi-num-2009-69-du-12-08-2009-jort-2009-065__2009065000691?shorten=iljB, accessed April 2016.

JORT (2006) La loi organique des communes modifiée et complétée par la loi organique n°2006 -48 du 17 juillet 2006, http://www.cnudst.rnrt.tn/jortsrc/2006/2006f/jo0592006. pdf, accessed May 2016.

JORT (2013) Décret No. 203236 du 2 Août 2013 portant modification du décret n° 98-1428 du 13 juillet 1998, relatif à la fixation du tarif des taxes que les collectivités locales sont autorisées à percevoir.

JORT (1998) Décret n° 98-1428 du 13 juillet 1998 relatif à la fixation du tarif des taxes que les collectivités locales sont autorisées à percevoir, http://www.cnudst.rnrt.tn/ jortsrc/1998/1998f/jo05998.pdf, accessed May 2016.

JORT du 24 juillet 1998, no. 59, http://www.legislation.tn/sites/default/files/fraction-journal-officiel/1998/1998F/059/TF199814283.pdf, accessed April 2016.

JORT (1977) Loi n° 11-1997 du 3 février 1997, portant promulgation du code de la fiscalité local (web copy not available).

JORT (1975) Loi n° 74-33 of 14 May 1975 Portant promulgation de la loi organique des communes de 14 mai 1975.

14 Street traders in post-revolution Cairo

Victims or villains?

Nezar A. Kafafy

Summary

Since the late 1980s, Cairo has witnessed a sharp reduction in formal jobs and an influx of workers into the informal economy. This chapter examines the changing power relations, and the role of street vending in these, at a unique moment in Cairo's history, and the power struggle between traders and the altering state.

Introduction

In December 2010, riots in Tunisia were sparked by the desperation of a street vendor, Mohamed Bouazizi, who set himself on fire after his goods were impounded and he was humiliated by the authorities. This horrific event sparked a public outcry for democratic reform in Tunisia that spread throughout North Africa and the Middle East, precipitating the downfall of presidents Ben Ali of Tunisia, Mubarak of Egypt and Gaddafi of Libya. Since the dramatic events that rocked these nations, their political and economic structures have nearly collapsed, and those currently in power continue to struggle with rebuilding them.

This chapter draws on research that took place in the midst of the political and economic upheaval of the Egyptian revolution. Egypt's revolution broke out on 25 January 2011, when anti-government protestors occupied Tahrir Square and other prominent locations (BBC, 2016). The occupation continued until 11 February 2011, when President Mubarak fled, to be replaced by an army council. Parliamentary elections over several months in 2011–12 gave power to Islamist parties and, in June 2012, the Muslim Brotherhood candidate, Mohamed Morsi, narrowly won the presidency. Our fieldwork took place in September 2012, during a brief period of relative political stability. However, by 2013, trust in the government was failing and, following widespread public demonstrations, the army stepped in to depose Morsi. In May 2014, former army chief General al-Sisi won the presidential election (BBC, 2016).

In pre-revolution Egypt, the phenomenon of street vendors was common and numbers and locations of vendors fluctuated, based on the intensity of police crack-downs and the economic environment. After the revolution, street vending flourished and was clearly visible on almost every street corner, impacting the everyday life of citizens. People debate whether vendors were victims or villains, and question whether the democratic reforms would solve street vendors' problems. After the democratic parliamentary and presidential elections in 2012, when government circles debated how to legitimate the informal economy, the deposition of the government in 2013 and subsequent street protests has led to a continuation of the former approach of crack-down and control, in which no accommodation has yet been reached.

The main driver, or root cause, of informal street vending is simply the need for income. The International Labour Organization's (ILO's) 15th International Conference of Labour Statisticians (1993, para 5(1)) defined the informal sector as 'units engaged in the production of goods and services with the primary objective of generating employment and incomes to the persons concerned' (ILO, 2013: 14). These activities operate within a small sector, with little division if any between labour and capital as factors of production. Labour relations are socially determined, rather than formalised with contracts and legal guarantees, although sometimes supported by documentation only recognised within the informal economy.

Economic informality is characterised through two domains: the lack of formal business registration, taxation or licensing documents; and the operation on public or private property that is not legally designated for the practice. Often, however, there is a continuum between formality and informality – for example, vendors may pay daily fees for space, but not be registered.

This study focuses on the commercial use of public space: squares, streets and station forecourts, and the mediation between the legal and policy frameworks of urban planning, land rights, roadway legislation and urban management to accommodate vending in public space. It is mostly in informally appropriated public spaces where police crack-down, harassment and brutality are a common response of those in power. The repression of street vending in Cairo is not unique to North Africa. Totalitarian and democratic governments alike have been antagonistic to the sector. A wide range of laws and regulations make it difficult for street vendors to work within the law. At the heart of the contest over public space is an unspoken belief that the 'untidy' poor and their 'anarchic' occupation of the street have no place in the global city, where the state and capitalist power are demonstrated through control of public space and the 'modernisation project' (Brown, 2006; Madanipour, 2010; Brown et al., 2010, 2015).

This research analyses information from a review of relevant legislation and bylaws; newspaper and grey literature on events preceding and following the revolution; publicly available statistics; interviews with 26 key informants in government and international agencies, the formal sector, academics, national and municipal officials, politicians and activists; and an extended questionnaires with 103 street vendors. Vendors were selected from four main market areas: Helwan (adjacent to Helwan Metro Station and its forecourt); Embaba (Embaba Tunnel and the slightly more up-market area of Ard El Gamiya); Demerdesh and Ramses (outside the metro stations, in front of public buildings and in nearby streets); and Tahrir Square (near Abdel Monem Riad micro-bus station).

The research took place during the brief period of democratic rule in 2012, and thus examines the change in power relations at a unique moment in the history of Cairo. Post-conflict cities where the urban economy has to be rebuilt are perhaps an extreme example of the wider impacts of structural adjustment. Post-neoliberal reforms, which have been transforming the economy of Egypt since the late 1980s, have been met with a sharp reduction in formal jobs, and have led to a shift of middle- and lower-income workers into micro-enterprises and informal economy work. Cairo, however, remains in conflict as the burgeoning growth of informal economy livelihoods met with a new recognition and dynamic attributed to the power of public voice that carries authentic claims for economic inclusion. However, as the post-2013 events have shown, the front lines of policing and security of the vendors is an ongoing battle that fluctuates, as those in power struggle with superficially remedying the symptoms while still being unable to address the reality of root causes. In the meantime, order and access to the urban public arena is a continual struggle.

The macro-economic and legal context of informal trade in Cairo

Since the mid-1990s the value added of manufacturing and agriculture, as a percentage of Egypt's GDP, has steadily declined. Services have been the largest growing sector of the economy, particularly commerce (Figure 14.1). Within the commercial sector, increasing disparity between imports and exports has meant that retail distribution of imported goods continues to be a key livelihood opportunity (Figure 14.2). Unemployment, which had dropped between 2006 and 2008, nearly doubled from 2008–11 and reached 12%. Consumer price inflation continued to rise faster than job opportunities. Accordingly, the Egyptian economy had been on an unsustainable trajectory to provide jobs for the existing labour force since well before the 2011 revolution.

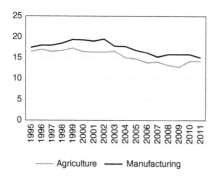

— Agriculture   — Manufacturing

Figure 14.1 Manufacturing and agriculture, value added by year (% contribution to GDP).

Source: TE, 2012.

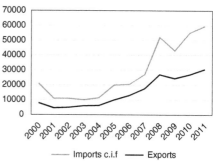

— Imports c.i.f   — Exports

Figure 14.2 Imports and exports 2000–11, Egypt (million US$, all partners).

Source: IMF Direction of Trade Statistics (October 2012).

These trends, exaggerated by political turmoil, clearly indicate structural macro-economic issues that cannot rapidly be corrected. Egypt is one of many countries whose workforce has responded to job losses associated with structural adjustment programmes with significant growth in the number of small and micro-enterprises (El-Mahdi and Rashed, 2009). GDP growth and investment trends have been accompanied by declining employment in the productive sectors. The decrease of manufacturing and formal service jobs in the overall economy is linked to global economic and trade policy and cannot be easily reversed at the local level. What needs to be addressed is the accommodation of viable livelihood strategies for the poor in the absence of other employment.

Writing in *The Wall Street Journal* in 2011, Hernando de Soto summarised key findings from his analysis, conducted in 2003, of Egypt's informal economy ('extra-legal economy' as he called it), presented to the Egyptian Cabinet in 2004. His research found that 26% worked in the public sector, 31% in the private sector and 43% in the 'extra-legal economy'. 'Egypt's underground economy was the nation's biggest employer. The legal private sector employed 6.8 million people and the public sector employed 5.9 million, while 9.6 million people worked in the extra-legal sector' (de Soto, 2011).

As far as real estate is concerned, 92% of Egyptians hold their property without formal legal title. De Soto estimated the value of all extra-legal businesses and property, rural as well as urban, to be US$248 billion – 30 times greater than the market value of the companies registered on the Cairo Stock Exchange and 55 times greater than the value of foreign direct investment (FDI) in Egypt since Napoleon invaded, including the financing of the Suez Canal and the Aswan Dam. (Those extra-legal assets would be worth more than US$400 billion in today's dollars) (ibid.).

UNDP's 2010 *Egypt Human Development Report* focused on the economic and social inclusion of the country's youth, and identified street vending as the second fastest-growing occupation among this sector of the population (from 2000–7), following construction work (UNDP, 2010). These occupations grew more than 5% per annum, adding more than 20,000 workers per year. Although the report noted that the growth of street vending had slowed during the second half of the period, it speculated that, as a long-term trend and along with some marginal farming occupations, street vending is likely to grow whenever unemployment rises, absorbing frustrated job seekers (ibid.: 160). The digital revolution, particularly mobile phones, is also influencing many other kinds of informal activity, such as street vending (UNDP, 2015).

Given both global and local uncertainty, it seems likely that unemployment in productive industries and tourism will take many years to address. In the meantime, self-generated livelihoods may indeed be the only short-term strategy to combat increasing poverty and deprivation (Al-Youn, 2012a). The macro-economic conditions in the country in general, and in Cairo in particular, provide the necessary stimuli for the development of informal economic activity as identified by the Egyptian Center for Economic Studies in a 2012 study (Abd el-Fattah, 2012: 3). This highly relevant study of 180 employees and 90 employers in micro-businesses in Manshiyat Nasr (a low-income area of Cairo known for waste recycling) noted that trade stands out from both manufacturing and services in its incentives for informal engagement. Difficulties of finding alternative jobs of informal trade workers were 80%, compared with 70% and 71% of employees in manufacturing and services, respectively.

In the same study, the legal framework for retailing and trade is seen as a more complex barrier than in other sectors, as 43% of employers attributed their preference for informal trade to the bureaucratic and legal difficulties involved in formalisation, compared with 30% and 27% of employers in manufacturing and services, respectively (ibid.: 11). This analysis suggests that legal and economic factors mean that street vendors are less likely than employees in informal services or manufacturing to have a realistic alternative livelihood.

Informal trade and urban public space

Like most developing and transitional countries, Egypt continues to urbanise, with Cairo being the principal magnet for rural–urban migration. Of Egypt's nearly 90 million residents, it is estimated that 22–25 million live in Greater Cairo (an increase from 16 million in 2000). Estimates suggest that nearly half live in informal settlements, over an area of about 243 square kilometres. In addition to the pressure on local labour markets from migration and population growth, there are some specific economic constraints. Post-revolution losses in FDI and the near collapse of tourism, particularly in Cairo, further reduced employment opportunities in the metropolitan area. Urban development on agricultural land along the Nile is driving up real estate while decreasing agriculture and work opportunities.

The tensions between the need for jobs and preservation of Cairo's historic fabric requires the resolution of certain conflicting interests: the National

Figure 14.3 Cairo: public space became both a symbol of the revolution (graffiti, left) and place of survival (right).

Organisation for Urban Harmony, concerned with the preservation of historic Cairo and its urban forms and traditions, has highlighted the importance of physical solutions which do not cause damage to historic buildings and building fronts, or compromise the 'dignity' of public spaces. The value it places on this is evident also in the design of projects which set out to accommodate vending without compromising these principles.

Starting from that premise, the question is how to resolve congestion of public space, rather than of demonising vendors and clearing them. Forced evictions and relocation to sites not commercially viable for vendors to use as has been tried on several occasions but failed. The original vending sites were mainly selected by vendors to reach pedestrians, but the new locations are not based on the same commercial logic. Relocation of vendors to designated locations in relatively deserted areas has ultimately led to their abandonment of the new venues and return to their old places. As a result, public funds are lost on deserted new markets.

Street vending and the law

The 2014 Constitution of the Arab Republic of Egypt, 2014, approved by referendum, sees work as a right and duty (Article 12) and will protect workers' rights (Article 13) with an economic system that aims to achieve prosperity through sustainable development and social justice that increases job opportunities and eliminates poverty (Article 27).

However, legislation concerning street vending relies on a law that is over 50 years old, the Peddlers/Street Vendors Law, No. 33 of 1957. This criminalises peddlers who trade without a licence, and prohibits vendors from selling on public transport, standing next to shops selling similar goods or in places identified by the police as needed for traffic or public security, or causing a disturbance. In Morsi's presidency period and in December 2012, Law 33 was amended by Law 105/2012, which increased the penalties to imprisonment for three months or a fine of LE 1,000 for the first offence, rising to six months or a fine of LE 5,000 for a second offence and confiscation of goods. This amendment was hotly contested by street vendors, who gathered outside five governorate buildings on Sunday, 6 January 2013, protesting against the law that toughens punishments for vendors, where the secretary general of the Street Vendors' Independent Union in Cairo, said, 'We call for freezing this law until our situation as street vendors is legalised'; the Street Vendors Union referred in their statement to the Sunday protests as a spark of the 'Hunger Revolution' (Mihaila, 2013).

Policy and debate over informal street vending in Cairo

Senior figures in most agencies understand the macro-economic conditions forcing the growth of street vending and accept that a sustainable solution for Cairo's streets should accommodate strategically the historically prevalent and increasing trend of street vending, but this understanding has not yet been implemented.

The UNDP estimates that at least 48% of trading passes through informal means, and the 2005 census data suggested the total number of informal sector workers was 7.9 million (UNDP, 2010). UN-Habitat considers that street vending provides employment for poor people, but previous governments continued to handle the issue in the same manner as their predecessors. It placed street vending as the cause of congestion, an untidy urban form and increasing security issues, rather than accepting that it is a symptom of the deeper-rooted economic issues. At one stage there was a move to implement one-day markets in Giza (i.e. weekly markets), but this does not provide the same financial returns for vendors as daily trading, and accordingly does nothing to address the issue. From 2014, the government seemed to have gained a better understanding of the issue, as it tried to provide more applicable realistic solutions for these vendors.

Key informant interviews revealed ambivalence in the understanding of the dynamics of street vending and its mitigation. There was consensus that the informal economy is critically important to the livelihoods of the poor, and that the way forward lies in incorporating and trying to fulfil the needs of the poor (including many street vendors) in any future policies and actions, although many view the negative impacts, such as the widespread encroachments of street vending on public space, as creating significant obstacles to legitimisation. The proposed steps to increasing legitimacy include training and education, inclusive and visionary planning and licensing.

Street vendors are frequently viewed as a major cause of traffic congestion. They do add to traffic congestion – they use pedestrian sidewalks and often encroach onto the street and parking areas – but statistics show that they are not the primary cause of congestion along highways and arterial routes. The Cairo Traffic Congestion Study Phase 1 (WB, 2010: xii) noted the high frequency of vehicle breakdowns, accidents, security checks, unauthorised micro-bus stops, overflowing bus stops and random pedestrian crossing are the key obstacles to free flow on arterial routes and highways. While certain notable spots on highways have more recently been used by vendors, this localised vending occurs in relatively few places.

Press articles suggest that local opinion of residents and merchants within residential and commercial areas is not favourable to vendors. There are complaints about overcrowding of pedestrian and vehicle routes, blocking streets, harassment of women, violence, drug dealing, reduction in property values in neighbouring areas and social status. 'These vendors take to the street, and they drag everything down to their level.... They're there and in your way, and as a resident, you can't do anything about it' (Mohsen, 2012). There is also resentment from formal shopkeepers. Key informant interviews support this conclusion, 'They (vendors) are colonising more space every day and crossing all boundaries, they ought to respect the space they are in, and actually they violate our privacy and humanity' (key informant interview), although evidence suggests that some shopkeepers are using street vendors to sell directly to pedestrians. In contrast, the vendors call for their rights: 'There's no justice – I go the legal route and I'm still punished. Give me my rights and I'll give you yours!' (vendor selling newspapers and soft drinks, Abdel Monem Riad bus station).

Another public debate associates the increase in street vending with an increase in harassment of women on the street, although it should be noted that such harassment is by no means new. In 2005 the Egyptian Center for Women's Rights (ECWR) launched a campaign to tackle sexual harassment in streets and public places. To document the status they released a report in 2008, based on interviews with 1,000 Egyptian women and 109 foreign women, as well as with 1,000 Egyptian men. Their sample suggested that 83% of Egyptian women and 98% of foreign women had experienced some form of harassment, while over 60% of Egyptian men admitted to harassing women, mostly verbally (Rizzo, 2012).

During the revolution, there was a significant reduction of harassment of women in Tahrir Square, although this would have been unthinkable only days before. Yet, shortly afterwards, harassment resumed and increased dramatically especially in the demonstrations in Tahrir Square in 2013; women were reportedly raped. For safety and political reasons, the government prevented demonstrations in Tahrir Square and closed the Tahrir Metro interchange for nearly two years. The station reopened in June 2015. This raises the question of whether a post-revolution society, in which livelihood opportunities are respected and informal livelihoods incorporated rather than harassed and repressed, might lead to a reduction of sexual aggression and the renegotiation of a better gender contract.

To whom should public space provide a canvas? Who should determine the identity, content, use and control of public space? This debate is eloquently summarised by Attia (2011), who argues that public space should be an organic expression of gradually changing social norms and attitudes. Our research suggests that there is a great deal of contestation within society over the control and role of public space. Neighbours, nearby businesses, passers-by and vendors have differing perspectives on the question. People interested primarily in the socio-political value of Tahrir Square, for example, have different objectives and demands from street vendors. In turn, this suggests that management proposals need to accommodate a wide range of needs and consult fully with stakeholders. The dissatisfaction of neighbouring residents and businesses are inflamed by the rather militant attitude adopted by Cairo and Giza governorates, whose approach seems to be that the provision of orderly solutions to the accommodation of vending can only come after a clean sweep of the streets. In August 2014, vendors were swept from Tahrir Square and central Cairo and, in early October 2015, from the area around the terminus at Helwan Metro Station – where more than 3,000 sellers were evicted.

According to Kafafy (2014: 14), Some of the municipal policies proposed are potentially problematic in terms of providing livelihood opportunities for vendors, such as calls for one-day markets (the idea that people close to the poverty line working more than 70 hours a week will make ends meet on that basis is unrealistic). A second example is in Giza Governorate, where the eviction of vendors by police resulted in outright clashes, injuries and fatalities (Al-Youm,

2012b). Another example involves the installation of municipal markets for registered vendors where they will not pose 'an obstacle'. However, this approach depends on the relocation of businesses to sufficiently pedestrian-rich areas, creating a significant planning challenge that would require stakeholder engagement to successfully accomplish (Mohsen, 2012).

Concurrently, the adoption of zero-tolerance policing has led to contestation. The family of the 'slain vendor' (Al Ahram, 2012) refused to bury him before the police gave an undertaking to prosecute the officer involved in his death. Egyptian vendors are not unionised in any formal sense. Nevertheless, they do have some collective power to resist the state and to negotiate the terms on which they are, or are not, accommodated. Thus it is important to discuss the future of street vending in Cairo, understand the nature of the phenomenon and its role in supporting the poor and their networks. Sustainable policies need to be devised, which allow the poor to survive and allow them and their children to participate in the jobs created by macro-economic policy in the country over the next few years.

Profile of vendors

Street vending in Cairo is dominated by men, but there are some women venders. Thus 93 of the vendors interviewed for this study were male, and ten female, all of whom were in Embaba, although a few were observed vending in the other market areas studied.

Very few respondents were under 18. The largest group was aged 19–34 (53%). Some 36% were aged 35–54 and only 9% of respondents were over 55. Nearly two-thirds were married (63%) and one third single (33%). Divorce or separation existed in only three cases among respondents, while one female was a widow and had been forced into vending by widowhood.

Education levels witnessed interesting findings as 28 respondents (27%) had been in education for more than 14 years, and six were university degree holders, while two had a masters' degree, one in international law and other in commerce. However, 39 respondents were illiterate (could neither read nor write) and were mostly migrants, often from Upper Egypt and other poor regions. Of the rest, 17 respondents had less than six years of education but could read and write; seven had six years of education (primary-level education); and 12 respondents (12%) had entered secondary schools and had 12 years of education.

Despite the small size of their businesses, 80% of respondents were main earners in their household and a majority were the sole earners (72%). They have dependents outside their household too; 37 vendors were born outside Cairo (first-generation migrants), of whom 22 sent remittances to rural areas. Cairo-born vendors also have financial responsibilities to relatives outside the city, with 29% sending remittances to relatives in rural areas.

The decision to start vending: do vendors have alternatives?

Vendors themselves believe that they have no viable livelihood alternatives. Some 21% of the 103 informal vendors interviewed had previously worked in a sales job; 41% had previously been employed, working for private or public enterprises which had shut down; and a few had lost their businesses as a direct result of the turmoil – for example, one merchant's warehouse had burned down. All those who had changed job claimed to have been involuntarily dismissed from their previous job. Some 28% were previously unemployed and nearly 8% were students. Overall this suggests that, as in many countries, informal street vending provides jobs on a large scale for people vulnerable to a weak job market, and has done so for a long time (Brown *et al.*, 2010; UNDP, 2010).

Management of vending space

Although the details vary from one market to another, the management of vending sites showed some commonalities. There is a direct relationship between individual vendors, the municipality and the police. A few vendors are licensed but they are a minority; those who are not licensed – the overwhelming majority – are vulnerable to enforcement or extortion. Vendors reported that most extortion is carried out by low-ranking police officers because the more senior officials change every three years, so they do not have a chance to build up complicated bribery networks.

In Helwan, vendors often pay 'controllers' or middle-men, often young men from noted families who negotiate between vendors and the police. There are two patterns: either, a vendor starts individually and, after working long enough, becomes a boss and calls relatives to join him, which increases his wealth and power; or vendors arrive as a group, one of which among them is the boss. The group supports him working under his supervision. These are often people from Upper Egypt who come as big families. Vendor interviews suggest that migrants tend to use this support system. In contrast, Cairo natives generally lack such strong support, as 'people in Cairo don't know each other' (male, selling fruit in Helwan).

Neighbourhood bonds can also form a basis for strong networks and social organisation among vendors. For example, vendors in a market area in Embaba voluntarily move their trading area out of the way of commuters during the morning rush hour, and at those times sell only to people who stop to shop. They also pay a weekly fee to their elected leader, who negotiates with police on behalf of the group. Vendors there say that there are unwritten laws which govern behaviour and that newcomers quickly learn and follow these unwritten laws. Vendors from every site reported that disputes among vendors are resolved with their leaders. Additionally, much of the responsibility for cleaning and organising is undertaken by the vendors themselves, by occasionally paying someone, or by tipping municipal cleaners.

These examples suggest that vendors are capable of drawing on a range of shared interests in negotiating with local stakeholders. Through dialogue and using appointed leaders, they develop management structures and negotiate the use of space and rights of way. Leaders of the vendor groups can represent the group in dialogue with municipal planners and others to tackle the perceived and real negative impacts of street vending.

Increasing number of vendors?

The number of newcomers to informal street vending sites in this study shows a clear increase over time. Figure 14.4 summarises the trend, showing the number of vendors beginning to trade in their current market area since one respondent's arrival in 1967. It is evident from the chart that the percentage of new arrivals has progressively increased over the years, and more significantly in 2011 and 2012. Although the figures do not show departures from the trade, the fact that many had been trading for many years suggests that often vendors do not leave the profession – for example, 40% of vendors interviewed had been trading for more than ten years, and a further 10% from five to ten years. New arrivals remain a relatively small proportion of respondents in the survey, although this could be different in other areas of the city.

The growth trends were generally similar across all four markets studied, although there were some local variations. The common pattern of higher influx of new vendors reflects the direct response to global and national economic conditions at the time.

Vendors were split on whether their vending areas had attracted many new vendors since the revolution. Most of the 103 respondents (59%) said 'many' had come. In contrast, 8% had noticed 'few' and 30% responded 'none'. Four others

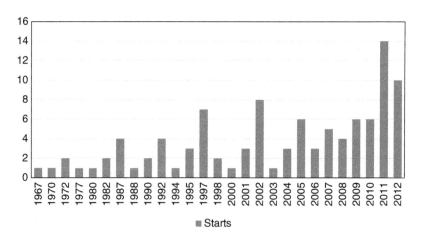

Figure 14.4 Vendors by year of arrival in current vending area.

Source: vendor interviews, 2012.

noted that '[new vendors had] tried but failed' to join their area, or that there had been only a temporary influx that later left.

Vending and the state

The revolution had a profound impact on the policing environment for street vending. Table 14.1 summarises vendor experiences of policing and order problems before and after the revolution. The general pattern of policing is consistent. For the brief interlude since the revolution and before the surveys were undertaken, police impingement on vending activities had radically decreased. For example, 76% of vendors said that they had experienced police harassment before the revolution, but only 33% following the revolution. Although the 76% may have been affected over a relatively long period, the answers are consistent across most aspects of vendors' experience of policing; in Table 14.1 experiences highlighted with an asterisk are significant at 5%. In qualitative answers it was also repeatedly stated that the police made far less frequent appearances on the street since the revolution.

The picture on civil-order conflicts, with nearby residents or other street vendors, seems to have remained the same or increased, although violence was reported to have drastically reduced. Violence had been experienced by 21% of respondents pre-revolution, falling to only 8% post-revolution (Table 14.1), despite the greater crowding among vendors and the reduced police presence.

From 2014 the tide began to turn as municipal authorities aimed to regain control of public space. The initial focus was on key public spaces such as Tahrir Square, mid-town Cairo, Helwan Metro Station and other spaces. The vendors were well aware of the significance of this drive, both short term and long term.

Table 14.1 Cairo: conflicts before and after the revolution

	Pre-2011		Post-2011	
Police conflicts	%	n	%	n
Harassment*	76	72	33	31
Confiscations*	65	62	9	9
Fines*	57	54	15	14
Evictions*	54	51	28	27
Prosecutions	14	13	1	1
Relocations*	100	95	1	1
Civil conflicts (non-police actions)				
Harassment	4	4	6	6
Confiscations	5	5	1	1
Violence	21	20	8	8
Thefts	5	5	9	9

N (100%) = 95.

Source: Extracted from vendor interviews, 17–25 September 2012.

Continuing conflict

After the fieldwork was complete, battles for the street continued. In December 2012, Charbel (2012) reported crack-down operations on street vendors in Cairo, Giza, Daqahlia, Mansoura and Alexandria. Proposals were discussed for relocating vendors to outlying satellite cities and vendors started to organise. Clearances were ongoing. Of the estimated 5 million street vendors in Egypt, 3 million were reported to be in Cairo; attempts were made to relocate 1,750 vendors from the city centre to a site at Torgoman and, in April 2015, bulldozers cleared 1,250 vendors from Ramses Square, but the street vendors claimed that the proposed relocation site at Ahmed Helmy, behind the railway station, was too small to accommodate them (Rios, 2015).

Conclusions

This study set out to investigate the influence of Egypt's revolution on the urban informal economy and street vending. Focusing on Cairo, the research followed key debates around street vending, identified its importance to the country's economy and the drivers for its growth, analysed the profiles of vendors and their businesses and analysed the impacts of revolution on business, attitudes and prospects. As in many other countries, in Egypt the growth of informal economy livelihoods in general, and street vending in particular, is not a response to changing policing patterns or idiosyncratic livelihood decisions, but rather a direct outcome of macro-economic trends and policy. Understanding it is therefore pivotal to understanding the opportunities and constraints governing the livelihoods of the poor. The key findings emerging from this study were as follows.

Macro-economic context

1) In Egypt, the industries of manufacturing and agriculture are growing more slowly than GDP;
2) since the revolution, tourism has suffered and the country has experienced capital flight;
3) unemployment remains widespread particularly among those with poor educational attainments; and
4) distribution of food and imported consumer goods is a growth sector.

The role of street vending

1) Most informal street vendors are the main earners of their household, and many are also the sole earners;
2) many remit money to their families in rural areas;
3) street vendors work long hours and some remain very poor;

4)   many vendors have no other employment options;
5)   many street vendors have low educational attainments, but several university graduates and post-graduates were found working as street vendors.

The dynamics of street vending

1)   Many street vendors have been in business in the same area since well before the revolution;
2)   newcomers tend to gather in one place and often work with relatives in the same area. The majority depend on friends, family, or other contacts for access to space;
3)   access to space is mediated through some form of patronage. Only 20% had simply settled in their spot;
4)   contest over space is largely managed informally among traders; and
5)   chiefs or bosses collect money from vendors, regulate disputes and mediate conflicts.

The continual growth of informal street vending clearly has implications for the city and its economy, for vendors' livelihoods, consumption of public space and circulation in the city. Efforts to manage the phenomenon within the context of the national transition to address the core issue of high unemployment and a weak economy are necessary. Police harassment has considerably reduced since the revolution, but is coming back in more visibly, particularly in landmark areas such as Tahrir Square. In the meantime, harassment from civil sources appears to have increased, indicating discontent among stakeholders and residents.

Understanding how street vending can be managed productively will help move forward these key issues of debate including: seeking effective ways to share of public space for pedestrians, vehicles and commercial exploitation; improving the management and maintenance of historic spaces and buildings; improving gender relations in public space, which appear to have significantly deteriorated again; and, perhaps most critically, redefining the relationship between the public and the police to improve the regulations affecting public space.

The eradication of informal street vending is not achievable, particularly in light of the ongoing situation in Egypt, which will take years, if not decades, to improve. In the short and medium term, developing sustainable planning and solutions which encompass formal and informal development, and integrating these, is the most practical approach. The model of collective management and collaboration that exists among street vendors is the key element with which authorities need to engage. Vendors' organisations could be expanded and formalised as a medium for the management of public space. The home-town bonds among the families and migrants among the vendors develop a strong group ethos, which may explain the drop in violence among them after the revolution, in spite of increased numbers and decreased police presence. It is this ethos that the authorities need to learn from when dealing with the phenomenon of informal street vending.

Acknowledgement

The research was funded under a British Academy Small Grant project entitled, Economic Inclusion and Political Change: Impact on Street Trading of Regime Change in Tunis and Cairo (Project SG111249).

References

Abd el-Fattah, M. (2012) A survey-based exploration of satisfaction and profitability in Egypt's informal sector, ECES (Egyptian Center for Economic Studies) Working Paper No. 169.

Al Ahram (2012) Family of slain vendor agree to bury body after police promise trial, 16 October, Ahram Online, http://english.ahram.org.eg/NewsContent/1/64/55785/Egypt/Politics-/Family-of-slain-street-vendor-agree-to-bury-body-a.aspx, accessed October 2012.

Al-Youn, A.M. (2012a) Govt: poverty rate increased to 25.2 percent of population, http://www.egyptindependent.com/news/govt-poverty-rate-increased-252-percent-population, accessed 13 October 2012.

Al-Youn, A.M. (2012b) Police shoot street vendor dead, http://www.egyptindependent.com/news/police-shoot-street-vendor-dead, accessed October 2012.

Attia, S. (2011) Rethinking public space in Cairo: the appropriated Tahrir Square, *Trialogue*, 108(1): 10–15.

BBC (2016) Egypt profile – timeline, BBC News, http://www.bbc.co.uk/news/world-africa-13315719, accessed July 2016.

Brown, A. (ed.) (2006) *Contested Space: Street Trading, Public Space and Livelihoods in Developing Cities*, Rugby: ITDG.

Brown, A., Lyons, M. and Dankoco, I. (2010) Street traders and the emerging spaces for urban voice and citizenship in African cities, *Urban Studies*, 47(3): 666–83.

Brown, A., Msoka, C. and Dankoco, I. (2015) A refugee in my own country: evictions or property rights in the urban informal economy?, *Urban Studies*, 52(12): 2234–49.

Charbel, J. (2012) Street vendors mull forming union to combat state crackdowns, http://www.egyptindependent.com/news/street-vendors-mull-forming-union-combat-state-crackdowns, accessed June 2015.

De Soto, H. (2011) Egypt's economic apartheid, *The Wall Street Journal*, 3 February, http://www.wsj.com/articles/SB10001424052748704358704576118683913032882, accessed April 2016.

El-Mahdi, A. and Rashed, A. (2009) The changing economic environment and the development of micro-and small-enterprises in Egypt in 2006, in Assaad, R. (ed.), *The Egyptian Labour Market Revisited*, Cairo: American University of Cairo Press: 87–112.

ILO (2013) *Measuring Informality: A Statistical Manual*, Geneva: ILO (International Labour Office) http://www.ilo.org/global/publications/ilo-bookstore/order-online/books/WCMS_222979/lang--en/index.htm, accessed April 2016.

Kafafy, N. (2014) Right to urban space in post-revolution Cairo: a study for street vending phenomenon, ARCHCAIRO6: Responsive Urbanism in Informal Areas, Towards a Regional Agenda for Habitat III, November, Cairo.

Madanipour, A. (ed.) (2010) *Whose Public Space? International Case Studies in Urban Design and Development?*, Abingdon: Routledge.

Mihaila, L. (2013) First spark of 'hunger revolution': street vendors hold protests at five different governorates calling for better conditions, *Daily News Egypt*, http://www.dailynewsegypt.com/2013/01/06/first-spark-of-hunger-revolution/, accessed October 2015.

Mohsen, A.A. (2012) Cairo's governor is attempting to remove street vendors – again, *Egypt Independent*, http://www.egyptindependent.com/news/cairo-s-governor-attempting-remove-street-vendors-%E2%80%94-again, accessed October 2012.

Rios, L. (2015) Street vendors expelled from downtown Cairo, *Al Monitor Egypt Pulse*, http://www.al-monitor.com/pulse/originals/2015/06/egypt-cairo-street-vendors-leave-fines-authorities-april.html#, accessed July 2015.

Rizzo, H. (2012) La création d'une communauté mobilisée pour les droits des femmes: la campagne contre le harcèlement sexuel en Égypte, *Égypte/Monde arabe*, Troisième série, Gouvernance locale dans le monde arabe et en Méditerranée: Quel rôle pour les femmes?, http://ema.revues.org/index3024.html, accessed October 2012.

TE (2012) Trading economics, www.tradingeconomics.com, accessed October 2012.

UNDP (2010) *Egypt Human Development Report 2010: Youth in Egypt, Building Our Future*, Cairo: UNDP (United Nations Development Programme) and the Institute of National Planning Egypt.

UNDP (2015) *Egypt Human Development Report 2015: Work for Human Development*, UNDP (United Nations Development Programme), http://report.hdr.undp.org/, accessed April 2016.

WB (2010) Egypt – Cairo traffic congestion study – Phase 1, WB (World Bank), http://documents.worldbank.org/curated/en/2010/11/16603168/egypt-cairo-traffic-congestion-study-phase-1, accessed April 2016.

Part III

Claiming 'Rebel Streets'

15 Emerging themes for the new legal order

Alison Brown

Coda

Street trading is a modern economic response to twenty-first-century trajectories of urban development and globalising market economies and, while some street trade remains a survivalist outlet for local produce and manufacture, much is linked to global supply chains. As the chapters in this book show, processes that govern street trade are informal and socially determined, ranging from 'land-lord' control of trading locations to casual invasion of space. However, the formal legal system in which street trading takes place matters. Who holds powers to evict and how these are applied are issues of critical concern to street traders.

For many street traders, a key focus in their struggle for rights and recognition – to a secure trading space, decent working conditions, fair taxes and social protection – is protection from legal exclusion and access to justice and the rule of law. They aim both to change and challenge punitive laws and to seek fairer implementation of regulations that manage street trade. Traders often lack the organisation, capacity and information to achieve change, yet where they have organised and been adequately supported by NGOs or legal resource centres, they have been successful in claiming rights through the courts.

As our research has shown, urban law contains historic and political legacies that create conflicting bodies of law and, to a large extent, criminalise the long-established activity of street trading. The strengthening of municipal regulation, and perception that street trading is 'untidy' and inappropriate to a modernist development agenda, has created a body of regulation that is almost always harmful to the livelihoods of the poor (Chapters 1 and 2).

Urban law governing street trading includes a plethora of legislation, bylaws and regulations that, for street traders, makes legal compliance impossible. Relevant legislation covers policing and public order; highways; urban planning; land use; registration and licensing; labour laws; public health codes; market regulations; municipal management; and many others. Bylaws on street trade tend to be narrowly focused (e.g. on hawkers) and prohibitive rather than enabling in intent, and some bylaws such as regulations on cart-pushers, or kiosk owners are relics of colonial control. While many jurisdictions promote formalisation of the informal economy, in many cities formalisation is impossible without major legal reform.

At best, conflicts are ignored and street trading is tolerated, but a crackdown can be triggered by political events such as election of a new mayor, hosting a mega-sporting event or major development proposals. For traders, a crucial problem is the unpredictability of municipal operations and policing, and lack of opportunity to negotiate problems. Traders often report that evictions were carried out without warning at night so that they could not salvage their goods, or with the use of force and violence. Municipal authorities sometimes outsource policing with limited oversight of contracted services – for example, for street clearances.

Accommodating street trading and resolving problems of congestion can be challenging, but until street traders are recognised as legitimate urban workers with collective rights to use public space, the scope for conflict-resolution is limited (Chapters 3–6). Privatisation of municipal services such as water, electricity and solid waste management can adversely affect traders, both because their needs are not addressed and because municipal tender requirements often exclude small-scale operators from bidding for privatised municipal service contracts, although traders' organisations are well placed to manage markets or collect solid waste in trading areas.

Legal conflicts

Comparative findings from the research suggest eight core areas of state legislation that influence street traders' rights.

Constitutional rights

Many constitutions explicitly recognise the right to work and right to a livelihood, which is helpful to counter legislation that makes street trade illegal. However, constitutions provide 'weak rights' which often have to be implemented through supplementary law (Chapter 3). Street traders have occasionally used constitutional rights to claim legitimacy, particularly the 'right to life'. In both the successful cases documented in this book, in Ahmedabad and Durban, traders have used constitutional rights as the basis of their legal challenge (Chapters 8 and 9).

Access to justice

Although access to justice is one of the four 'pillars' of reform in the Commission for Legal Empowerment of the Poor's report (UNDP, 2008), in practice it is rarely achieved (Chapter 2). For street traders, barriers to justice include uncertainty of outcome, difficulties in understanding their rights, the costs in time and money and limitations on rights given their often 'illegal' use of space. However, where street traders have an effective organisation, and support from a *pro bono* advocate or NGO, they have been successful – as in Ahmedabad or Durban.

Urban space

Town planning regulations are a frequent justification for street trader evictions, and urban renewal and beautification projects often displace traders (Chapter 5). The development of gated communities with private streets and construction of purpose-built vending malls also cause problems for traders – for example, the *Machinga* (hawker) complex in Dar es Salaam remained empty years after completion as it was isolated from customers (Chapter 12). Street trading areas are rarely shown on planning maps because of prejudices that they should not exist or difficulties of mapping areas where boundaries are unclear. Here, the Indian concept of 'natural markets' as places where a significant pedestrian flow creates commercial opportunities for street trading is an important innovation (Chapter 8).

Property rights

Land rights may be considered as 'bundles' of rights which can include rights to occupy, use, transfer, inherit, sell or develop land. Public spaces – where much street trading takes place – are an important component of urban land, often covered by informal access rights, but these are rarely considered in urban land and property law (Chapters 5 and 6). Understanding and codifying access rights to public space is key. Customary land rights are common in much of Africa – many traders have an expectation that they should have rights to use urban land to survive, which, although not articulated as customary rights, frames a similar concept, as in Dakar (Chapter 11).

Highway regulations

Legislation often prohibits structures or activities that obstruct vehicle or traffic movement, privileging vehicle users and drivers over other highway users. Some jurisdictions permit the use of highways land for hawking, but this is rare. In Sénégal, the 1976 *Loi sur l'occupation de la rue* expressly forbids informal use of space (*occupation anarchique de la rue*) (Chapter 11). Although issues of traffic safety are crucial, many road reserves have the potential to accommodate other uses – for example, roadside trading or urban plant nurseries. Transport termini – bus termini, metro stops and train stations – attract considerable pedestrian flows and are a natural location for street trading, which is rarely recognised in urban transport design, as new Bus Rapid Transit Routes displace traders (as in both Ahmedabad and Dar es Salaam) and new bus termini offer few trading facilities.

Licensing and business regulation

In many cities few small and micro-businesses are licensed, but there is growing pressure from governments and international agencies to maximise government revenue and promote formalisation (Chapter 2). Registered businesses are liable for tax, although a distinct category may allow small businesses to register for simplified licensing and taxation. There has also been a drive to license specified occupations such as the sale of meat and fish. Contrary to popular view, many street traders pay extensively to use street space in daily fees or bribes to city officials or gatekeepers, but achieve little security. Formalisation increases security for traders, but, until regulation is appropriate and licensing criteria achievable, it will be beyond the reach of many.

Labour law

Informal economy workers are often excluded from the protection of labour law, which focuses on employees and often excludes the self-employed, or unregistered enterprises. Core elements of labour law which are relevant to street traders include employment contracts, occupational health and safety, social protection including sickness and pension benefits, rights to unionise and to engage in collective bargaining, and to the avoidance of discrimination at work (Chapter 7). Street traders often work in vulnerable situations, lacking protection from the weather, basic services and secure employment, which heightens the vulnerability of an insecure profession.

Urban management

The day-to-day operation of street trading is often covered by municipal and urban management bylaws that affect street trading – for example, highways and traffic control; solid waste management; market management; public health regulation (e.g. sale of cooked food); fees and licensing; and managing public parks. However, bylaws are often out of date and poorly coordinated across municipal departments. The powers of eviction, fines and court procedures are critical and are usually implemented by municipal or police authorities under powers to ensure public order and restrict loitering. The avoidance of summary evictions is a critical step in establishing legal rights (Chapter 12).

Claiming 'rebel' streets

Street traders use five main strategies for claiming 'rebel streets': covert occupation; direct action; political influence and bargaining; organising and unionisation; and legal challenge.

- *Covert occupation* involves sheer force of numbers and avoidance of direct confrontation with the authorities. Many street traders carry limited goods, and can move quickly if a patrol passes by, but this approach does little to address the inherent injustice of much legislation involving street trade.

- *Direct action* remains an unreliable route to reform. In Egypt the revolution has largely failed (Chapter 14). In Tunisia, demonstrations in sympathy with a street trader's plight sparked revolution, and the powerful quartet of trades unions and lawyers negotiated a peaceful regime change, but this does not yet benefit street traders (Chapter 13).
- *Political influence and bargaining* can create short-term gains, but can easily be reversed. The approach was partially successful in Sénégal, where traders' riots aroused political support and enabled newly formed trader organisations to negotiate with the municipal authorities (Chapter 11). In Dar es Salaam, political gains were short lived and trading remains illegal across multiple domains (Chapter 12). In China, local authorities themselves have led accommodation in order to balance the National Sanitary City campaign with the national discourse of social harmony (Chapter 10).
- *Organising and unionisation* is a key strategy for institutionalising gains, but traders need to build capacity and their knowledge of legal systems in struggling for their rights. Once organised, traders can more effectively lobby – both for the fair implementation of existing legislation and to change and challenge inequitable aspects of the law. Both approaches were effective in India.
- *Legal challenge* remains a relatively rare strategy, pursued successfully only in Ahmedabad and Durban after years of struggle. Here WIEGO's Programme on Law and Informality holds the promise of more wide-reaching and fundamental reform, leading to effective implementation of enabling local regulations, national legal reform and international recognition of worker rights as in the ILO conventions (Chapter 7).

Empowering trader associations through capacity building and international networking is crucial to conferring legitimacy and demonstrating the potential for enlightened urban law. The Street Vendors (Protection of Livelihood and Regulation of Street Vending) Act, 2014, in India is highly significant in that its intent is to reduce poverty by enabling the regulation of street vending – and to give street vendors influence over the implementation of its provisions. The law was promoted by street vendors, spearheaded by the National Association of Street Vendors of India (NASVI) to protect their rights (Chapter 8). This approach is in complete contrast to much of the legislation reviewed in this book, which is intended to repress and restrict the 'nuisance' of street trade.

In Latin America, the social contract between states and communities and the 'right to the city' agendas of Brazil and Ecuador have brought about reform of urban law and urban planning, leading to the inclusion of marginalised communities. Here both legal reform and pro-poor municipal policy has allowed street traders to claim urban space (Chapters 4 and 5).

An interesting aspect of the research is the different attitudes to the law in the countries studied. In India, the population has a huge respect for judicial processes as street traders lobbied at federal level for legal reform and challenged state governments in the courts over its implementation. In South Africa, traders also using the courts to claim legitimacy. Elsewhere in sub-Saharan Africa, where informality is widespread and local government is under-resourced, formal legal systems are taken less seriously and directly political lobbying is seen to be more effective than legal challenge. In China, a strict regulatory regime has been tempered by a national discourse on social harmony, leading to ambivalent policy towards street traders.

It is also clear from the research that, in situations of political change or civil unrest, the informal economy is acutely affected by instability and street traders and others must make rapid adjustments. In Dakar, a change in president and policy meant a shift from tolerance to control (Chapter 11). In North Africa, in the wake of the Arab revolutions, the collapse in the economies of Tunis and Cairo (Chapters 13 and 14) resulted in a rapid increase in informal work. The initial lifting of state repression and loss of authority for state institutions meant a brief window in which street trading was unpoliced, although power struggles over space emerged. As state authority was imposed again, street clearances once again became commonplace.

The core argument of this book is that, in the tensions between legal instruments as expressions of state control, and the needs of the working poor for urban space, are largely unrecognised and unresolved. The research community represented here argues for a reconceptualisation of legal instruments to provide a rights-based framework for urban work which recognises the legitimacy of urban informal economies, the scope for collective management of urban resources and the social value of public space as a site for urban livelihoods.

Rethinking law and rights for the informal economy

Five frameworks were posited at the start of this book as a basis for better understanding of law and rights in the informal economy. Each is discussed briefly below.

Legal pluralism: It is clear from this book that the legislative frameworks affecting street trade are complex and contradictory. As outlined in Chapter 2, the regulation of street trade operates across five domains of complexity: state law and bylaws; international conventions and law; local approaches to implementation; politics, and the informal norms and practices of street trade. Formal state law and bylaws often contain clauses and contradictions that make it impossible for street traders to achieve compliance. Addressing this challenge means accepting that trade is legitimate urban work, undertaking a legal audit, addressing conflicts in the law, and establishing *appropriate regulations* that seek to enable rather than criminalise street trade. Once these are established, collective systems of management have the basis for improving relations with local government and strengthening the livelihoods and working conditions of the poor.

Human rights: The accounts in this book make it clear that the discourse of human rights has so far been insufficient to safeguard informal economy workers from injustice and disadvantage (Chapter 3). 'Programmatic' constitutional rights have proven weak in protecting and furthering their interests. While there have been some small gains in asking courts to enforce the abstract rights in international instruments and national constitutions, for street traders pursuing progress through enabling domestic legislation which establishes specific, enforceable legal rights should be a priority. International comparisons of street vending legislation (Box 1.2) provided a useful basis for street vendors in India to lobby for new legislation (Chapter 8), but even implementation was by no means straightforward.

Right to the city: In Latin America, the social contract between states and communities and the 'right to the city' agendas of Brazil and Ecuador, has led to reform of urban law and urban planning, leading to the inclusion of marginalised communities. The concept has not been widely applied to street trading and the informal economy, but the 'right to the city' approach has considerable approach through both legal reform, and pro-poor municipal policy to allowed street traders a legitimate share in urban space (Chapter 4).

Rights to access public space: As outlined in Chapter 5, the detail of regulation and practice of urban management and policing combine to provide a powerful system of control over public space with profound implications for street traders. In some cities, supportive policy and practice is emerging alongside existing control. Allowing street traders managed access to busy city centres through inclusive and participatory negotiations are essential to enable the economic potential of public space and improved working conditions to be achieved, as the examples of Durban and Guangzhou show (Chapters 9 and 10).

Land and property rights: The on-going persecution of street traders in public space suggests an urgent need for the definition of a new 'rights regime', that sees the public domain as part of the *collective land resource* of cities, to which collective rights pertain (Chapter 6). Drawing on the work of Elinor Ostrom, public space is probably best described as a 'common pool' resource with open access, but common ownership and management. This would include a variety of rights – to work, move through and enjoy public space. For street traders, bundles of rights might include access and beneficial use, but with conditions requiring contributions to the collective management of the resource. In this, as in all aspects of legal empowerment, the effective organisation of street traders is key.

The chapters argue that an inclusive legal order that embraces street trade needs six key elements:

1) acceptance of the legitimacy of street trade;
2) empowered street traders' organisations and an effective negotiating platform with municipal government;
3) a legal audit to examine legislation that affects street traders and their trade;

4) advocacy on national legal reform to create enabling legislation to manage street trade;
5) campaigns to implement fair legislation, giving street traders powers in its implementation; and
6) innovative political and administrative leadership to negotiate change.

Only once these are in place can the economic and employment potential of street trade be realised.

Reference

UNDP (2008) *Making the Law Work for Everyone: Volume 2, Report of the Commission on Legal Empowerment of the Poor*, New York: UNDP.

Index

Taylor & Francis eBooks

Helping you to choose the right eBooks for your Library

Add Routledge titles to your library's digital collection today. Taylor and Francis ebooks contains over 50,000 titles in the Humanities, Social Sciences, Behavioural Sciences, Built Environment and Law.

Choose from a range of subject packages or create your own!

Benefits for you

» Free MARC records
» COUNTER-compliant usage statistics
» Flexible purchase and pricing options
» All titles DRM-free.

Benefits for your user

» Off-site, anytime access via Athens or referring URL
» Print or copy pages or chapters
» Full content search
» Bookmark, highlight and annotate text
» Access to thousands of pages of quality research at the click of a button.

REQUEST YOUR FREE INSTITUTIONAL TRIAL TODAY

Free Trials Available
We offer free trials to qualifying academic, corporate and government customers.

eCollections – Choose from over 30 subject eCollections, including:

Archaeology	Language Learning
Architecture	Law
Asian Studies	Literature
Business & Management	Media & Communication
Classical Studies	Middle East Studies
Construction	Music
Creative & Media Arts	Philosophy
Criminology & Criminal Justice	Planning
Economics	Politics
Education	Psychology & Mental Health
Energy	Religion
Engineering	Security
English Language & Linguistics	Social Work
Environment & Sustainability	Sociology
Geography	Sport
Health Studies	Theatre & Performance
History	Tourism, Hospitality & Events

For more information, pricing enquiries or to order a free trial, please contact your local sales team:
www.tandfebooks.com/page/sales

Routledge
Taylor & Francis Group

The home of
Routledge books

www.tandfebooks.com

Milton Keynes UK
Ingram Content Group UK Ltd.
UKHW040108071024
449327UK00019B/902

9 780367 138752